ESSAYS AFTER WITTGENSTEIN

Written within the tradition of Wittgenstein's work, these eight original essays in philosophical psychology are either by-products of efforts to understand Wittgenstein's later writings or applications of techniques and approaches derived from Wittgenstein to problems about which he did not say a great deal.

In much of his later writings, Wittgenstein was not so much trying to explain his own views as to tease, annoy, and confuse the reader into working out for himself solutions to some philosophical problems. Professor Hunter, goaded and guided by Wittgenstein, here presents in clear and plain prose the views that he has worked out on a number of different questions. Although the essays are not exegetical in form, they will be found by students of the great philosopher to contain a large number of novel suggestions as to how Wittgenstein might be interpreted; philosophers, psychologists, linguists, and mathematicians are offered an unconventional, interesting, and richly argued approach to some of the main problems in philosophical psychology.

The essays treat Meaning, Telling, Pain, Logical Compulsion, Identity, Imagining, Dreaming, and Talking. One eminent scholar has predicted that this volume may reverse the present tendency of philosophers to follow the lead of Noam Chomsky in the philosophy of language.

J.F.M. HUNTER is an Associate Professor in the Department of Philosophy at the University of Toronto.

Essays
after
Wittgenstein

J.F.M. HUNTER

UNIVERSITY OF TORONTO PRESS

©University of Toronto Press 1973
Toronto and Buffalo
Reprinted in paperback 2017
ISBN 978-0-8020-5278-0 (cloth)
ISBN 978-1-4875-9189-2 (paper)
LC 71-190345

Preface

The essays in this volume are either by-products of efforts to understand Wittgenstein's later writings, or applications of techniques and approaches derived from Wittgenstein to problems about which he did not say a great deal.

Wittgenstein asks a great many questions that he does not answer, suggests quite a number of experiments without also suggesting what results they will yield, makes cryptic pronouncements, and in general leaves one, at all too many junctures, extremely uncertain as to what point he is making. There are many ways of removing these uncertainties, but a method I have found particularly fruitful is that of tackling the problems he is discussing myself: trying to think a problem through until I find some way of looking at it that on the one hand can be set out as an arguable position in its own right, and on the other can without undue strain be defended as an interpretation of, or as a clue to, some of Wittgenstein's puzzling remarks on that and related topics.

To do this with any success one needs to be of a somewhat Wittgensteinian turn of mind; and even given that condition the procedure will not generally yield quick results. Many of the approaches one tries will either prove defective in themselves or will not check out as interpretations of Wittgenstein. Sometimes, however, the pieces seem to fall nicely into place; and at such times I have found again and again both that I had been away off in my first hunches as to what point Wittgenstein was making, and that the point I was finally disposed to attribute to him was in its own right both extremely interesting and capable of being taken very seriously.

At such junctures there arises the problem of whether to present the views reached in this manner as my own or as Wittgenstein's. In these essays I have largely taken the former course; and although this may count as intellectual larceny, it is recommended by the following considerations:

I The views I am expressing are generally complicated; hence it is difficult enough to present them clearly and as an attractive philosophical position without encumbering the presentation with a great deal of close textual analysis.

II The question whether what I have to say represents a correct interpretation of Wittgenstein has generally seemed less interesting than the question of whether it makes a contribution to solving or dissolving some old (and some new) philosophical problems.

III As an inevitable by-product of the method I was following, I often found myself with more to say about a question than there is any serious textual warrant for attributing to Wittgenstein; and usually it seemed preferable to go on freely rounding out a discussion or strengthening an argument, than to confine myself to what seemed to be demonstrably Wittgensteinian.

I offer these essays, therefore, primarily as independent and I hope interesting treatments of a number of topics; but I think they may also to some extent be used as a source of suggestions as to how Wittgenstein may be understood.

In the latter capacity, while much of what the essays contain may be familiar enough to people who read Wittgenstein, and may at best serve as a fresh or as a particularly clear presentation of what is well known but hard to explain, I think in quite a number of ways the Wittgenstein I depict is someone new to the philosophical public; someone, also, who deserves to be taken a good deal more seriously than the stereotyped Wittgenstein with his repertory of dark sayings about language-games, family resemblances, depth grammar, private languages, and the meanings of words being their use.

The essay called 'How do you mean?' seems to me to be predominantly only a fresh presentation of the familiar. Those essays that, I believe, cast the greatest amount of new light on Wittgenstein are 'Telling,' 'The concept of pain,' and 'Logical compulsion.' Those in which the tendency to go on in my own way is most marked are 'Personal identity,' 'Imagining,' 'Some questions about dreaming,' and 'On how we talk.' In any of the essays, however, if I am right, one may find all of these properties.

There is a certain amount of, perhaps too much, repetition of themes. The question how we know what to say, for example, turns up in the essays on Meaning, Talking, Telling, Pain, and Logical compulsion. If this proves tedious I must beg the reader's forgiveness. The good sense of eliminating such repetition by editing and cross-reference seemed to me to be counterbalanced by three main considerations. In the first place, while the same theme may be repeated, there is, I believe, very little repetition of arguments, with the result that something new is contributed by each appearance of any theme. Second, it seemed to me important to bring out the key role that certain notions do play in our thinking about a variety of questions, and that this could not be done as effectively by merely mentioning a theme at the various junctures where it is relevant as by spelling it out. Third, in view of the likelihood that essays in the volume will be studied individually, it seemed to me advisable to make each one as self-contained as possible.

I wish to express my gratitude to the University of Toronto for a leave of absence during which all but two of these essays were written; to the Canada Council for a Leave Fellowship; to the Department of Philosophy of Stanford University for their hospitality to me as a visiting scholar in the academic year 1968-9; and to the Department of Philosophy of San Francisco State College for the use of an office in the summer of 1969.

John Woods has generously read and commented extensively on nearly all of this material, and has goaded me at many junctures into probing more deeply than I would otherwise have done. The meticulous comments of my friend John Stratton have resulted in a very much lower level of muddle and inelegance than would otherwise have prevailed; and I have profited a great deal also from the criticisms of David Gallop, Rupert Buchanan, and Jack Canfield, and of the anonymous readers for the University of Toronto Press and the Humanities Research Council. Professor Norman Malcolm's comments on an early version of the essay 'On how we talk' were very useful to me. The manuscript is published with the help of a grant from the Humanities Research Council of Canada, using funds provided by the Canada Council, and a grant from the Publications Fund of University of Toronto Press.

I suppose I should thank some dear friends, an anonymous reader and the editors of the Press, for dissuading me from calling this 'The Pink and Mauve Book,' a name to which I confess some lingering partiality.

The following abbreviations have been used:

BB *The Blue and Brown Books*, Basil Blackwell, Oxford, 1958
PI *Philosophical Investigations*, Basil Blackwell, Oxford, 1953
RFM *Remarks on the Foundations of Mathematics*, Basil Blackwell, Oxford, 1964
Z *Zettel*, Basil Blackwell, Oxford, 1967

Contents

ESSAYS AFTER WITTGENSTEIN

How
do you
mean?

The question I am going to discuss is whether, when a person means something, for example when he says something and means it, or when he means this or that by what he says, or when he inadvertently says one thing although he means to say another, the meaning of it is something which goes on in him, an activity perhaps, or a process, a state, or an event. If it is one of these things it will more likely than not be a mental something or other; but I do not wish to exclude the possibility that it might be something physical, such as looking stern when he makes a dire threat and means it, or perhaps an increase in the pulse rate. So while I will mostly talk about the possibility that meaning is a mental phenomenon, my remarks will apply equally well to the possibility that it is something physical.

'How do you mean?,' the question I have given as the title of this essay, is therefore only one of the questions I shall be considering, and even so it has to be taken in a special and, you may think, peculiar way: that is as meaning 'What do you do *in order* to mean something?' or 'How do you

This essay originated as a lecture in the University College Public Lecture series for 1967–8, and was published in the *University of Toronto Quarterly*, xxxviii, 1, 1968. It appears here with various revisions. Although it was written independently, the essay might pass as a translation into continuous and not so cryptic prose of Wittgenstein's scattered remarks on these questions in the *Philosophical Investigations*. See, for example, Part I, the footnotes on pages 18, 33, 53, and 54; and §§ 138–9, 317–19, 334–8, 363, 378–81, 414, 431–3, 436, 449–51, 455–7, 501–14, 546–8, 589–90, 593, 601–6, 633–8, 648–93; and Part II, pages 175–6, 181–2, 214–19.

go about meaning something?' These questions presuppose that meaning is something we do; but it might also be something that happens in us, and if so we could call it an event if it happened briefly and intermittently, a process if it took some time and was marked by internal complexities, or a state if, like health, it involved a set of things being true of us at the same time.

The ensuing discussion may not prove of much interest to anyone who lacks all inclination to think that meaning something is something that we do, or that happens in us, and therefore it may be best if I begin by offering some positive considerations on this point.

First of all, many of the ways we express ourselves strongly suggest that meaning something is a distinct phenomenon, alongside of saying something. We say 'I said it and I meant it' and this looks like two things, not one; and when we say 'I said it without meaning it,' or 'I said it but didn't mean it,' this looks like a report of the absence of one of these things, which would normally be present. We say 'What I meant when I said that was ...' and this looks like a report of something that went on alongside the saying of something; we talk of 'trying to say what we mean,' and this looks like a report of the existence of a state of affairs called 'meaning something' prior to the existence of a state of affairs called 'expressing it in words.' And so on.

Secondly, when we use the verb 'to mean' in these ways, there does seem to be an inner state of ours that is importantly connected with our expressing ourselves the way we do. If we make a dire threat and mean it, we may be conscious of a feeling of grim determination. Or if we refer in conversation to Mr N and mean the tall man over there in the corner, we perhaps think of him, or look his way. And when we are struggling to say what we mean, there is a familiar if hard-to-describe experience of the imminence of successful expression which could be thought to be a pre-verbal knowledge of what we mean, or what we want to say.

Thirdly, it is generally either true or false that we mean a dire threat, or mean this rather than that by something we have said; and a natural, if not a necessary, answer to the question what makes it true or false is that it is the presence or absence in the speaker of an appropriate mental or other state. It is true perhaps that I mean a threat if I feel something like grim determination, but false if instead I merely feel amusement at the reaction of the threatened person.

While I have set out these reasons for thinking that meaning is something mental, I am inclined to think it is generally not because of sober consideration of reasons that we are likely to believe such things: there is much more conspiring to make us believe them than can ever be laid out succinctly in the form of reasons. We tend to be more convinced than a short list of reasons would ever warrant; and disposing of any given set of reasons is likely to leave us with a backlog of undiscussed worries that will generally be sufficient to keep us trapped. For this reason the kind of discussion I am about to embark on could aptly be called logical therapy – by which is meant not that people on whom the therapy is worked are treated as being sick, and their views treated as symptoms of something quite unlike the sensible and rational phenomenon they purport to be, but only that the discussion is likely to range into all kinds of miscellaneous corners of human thought, and is likely to consist more often in displaying ways of getting started on the jobs to be done than in actually covering all the vast territory to be covered.

I am proposing to show – or show how it can be shown – that meaning something is neither a mental nor a physical phenomenon; but before embarking on this project I had better, if only to keep everything philosophically pure, say something about how various are the uses of the verb 'to mean.'

To begin with, there are uses having nothing to do with this issue, for example where certain states of affairs mean, that is to say indicate something. A red sky in the evening is said to mean fair weather tomorrow, and certain medical symptoms mean that a person has a certain disease. 'Means' is also sometimes about equivalent to 'matters' or 'is important,' as in 'It means a great deal to me whether he comes.' Sometimes also it is about equivalent to 'intend,' as in 'I have been meaning to write for a long time.' And much the commonest of the uses we are not discussing are those connected with the asking and giving of the meanings of words.

The class of uses we are discussing can be roughly defined by two characteristics: first, a person is the subject of the verb 'to mean,' and second (to exclude cases like 'meaning to write'), his meaning something is directly tied up with his saying something.

One might think that for the class so defined (or indicated) 'mean' always meant about the same thing, but in fact there are considerable differences. When we correct a slip of the tongue or other speech ineptitude, saying 'what I meant to say was ...,' we are scrubbing one utterance and replacing it with another; when we interpret something we have said, saying 'what I meant by that was ...,' we are giving an alternative rendition of what we have said; when we explain ourselves further, saying 'what I mean is ...,' we are not explaining the meaning of what we have just said, but simply saying more along the same lines; when we say something and add that 'we mean it,' we are not further explaining what we have said, but indicating that it should be taken just as it stands; when we say that we meant something as a joke or as a compliment, we are not straightening out or clarifying what was said, but so to speak indicating the auspices under which it was said; and when we struggle to express ourselves and call it 'trying to say what we mean,' there is nothing that we are correcting, explaining, or adding to.

You might say that this does not show that 'mean' has various meanings, but only that different sentences in which it appears mean different things: the word itself always means about the same. We could substitute 'intend' for 'mean' in all these sentences, and they would have the same meaning. And I would not deny that this substitution could be worked. But it seems to me only to obscure the differences of sense – which lurk equally if not more in the use of the word 'intend.' If I say 'It is hot' when I meant (intended) to say 'It is humid,' it is not generally true that I formed the intention of saying 'It is humid,' but something slipped and I said 'It is hot': I just see now that 'humid' would have been a better word. And if I explain what I meant by something I have said, it is not generally true that I had formed an intention of saying what I now offer as elucidation, but for some reason instead made the remark that I am now elucidating: it is just that it now seems to me that I can clear the air by this way of elucidating my original remark.

Yet it may not matter whether the differences I have noted are differences of meaning or of something else: in any case there are differences, and I shall have to be careful not to overlook them. There will not be a great deal I can say in general as to whether meaning is something mental (or physical), but there are some general points, and I will begin with a few of these.

If meaning is something that goes on in a person, it is probably true that it must be a process, a state, an activity, or an event; and hence if we could show that it was none of these, then in the absence of other alternatives we would have a very general reason for doubting that it was any kind of phenomenon at all.

I suggest the following as a method of deciding whether meaning is any of these things: for any one of them, say an activity, decide on a few things that are activities if anything is. It will not matter, I think, whether you choose mental or physical activities, but for safety's sake, choose some of each. Then using these specific examples as guides, make a list of things which are sometimes true of activities. It will not matter how frequently these things are true; what matters is whether it sounds odd to inquire in any particular case whether this or that item on the list applies. Then, going on the principle that if anything is an activity, at least a fair number of these things which may be true of activities in general will be true of meaning, we can decide whether it is an activity by seeing how far it shares the characteristics of activities in general.[1]

Following this prescription, we might take as our stock cases of activities, thinking, writing a letter, or playing tennis. And some of the things that are sometimes true of these are that a person can be ordered to perform them, and decide, refuse, omit, or forget to do so. A person can be skilled or incompetent at them, can be busy doing them, be interrupted while doing them, and can find the performance of them easy or difficult, tiring, pleasurable, or boring.

If we now apply this list of things characteristic of activities in general to the case of meaning, we find that although there are senses in which we can order, or at least suggest, that a person mean this or that, they are not the ordinary sense, the sense in which a person can say 'Right!' and thereupon mean it: and we can't make a false promise by cunningly omitting to mean it, or a true promise by performing the act of meaning it as well as the act of saying it; we can't decline to explain what we meant on the ground that we simply forgot to mean anything; it is not skill at meaning things which enables us to mean one thing by another, or want of skill which prevents us from meaning false or meaningless things; it is neither pleasurable nor boring, tiring nor a cinch to mean by 'Mr N,' 'the tall man over there in the corner'; we can't be busy meaning something, nor can we be interrupted by a phone call in the middle of it and later have difficulty

---

1 This is a method often used, but not often described. It was used by Aristotle, for example in *Nicomachean Ethics*, 1173a31 *seq.* and 1174a13 *seq.*; by Hume, for example in the *Treatise of Human Nature*, Vol. I, Part IV, Sec. v; and by Ryle in places too numerous to mention. In cases in which the method turns up a very mixed bag of similarities and dissimilarities, it would be futile to try to refine it in such a way that it would still yield a decision – that is, to try to say 'where the line should be drawn,' or how many dissimilarities is too many. In those cases one may simply have to look for another procedure. But in many cases, including those that follow, it does yield a clear result.

remembering where we left off. All of which makes it extremely hard to see how meaning could be an activity, mental or otherwise.

Applying the method to processes, we might take as our stock cases learning and physical growth. These phenomena obviously share a durational characteristic with activities, and therefore many of the same things about being pleasurable or boring, being interrupted and resuming can be said about them, the main difference here being that where in the case of activities we would talk of doing, omitting, and forgetting, in the case of processes we would talk of wishing and hoping that the process would occur or resume, of noticing that it was going on, of its failing to occur or resume, and of being delighted or disappointed that it should or should not occur. Processes also are typically marked by stages and developments, and move towards some sort of completion.

But again none of this applies to our meaning things. We do not wish that we would mean by 'Mr N,' 'the short man with the beady eyes,' or notice with surprise or with dismay that we are well on the way towards meaning 'the man with the quaint sense of humour'; we would not know what to make of the question how long it took to mean something, or whether it went through more quickly when we were in good health, nor could we suggest any typical stages or developments in the process of meaning something.

If we were to apply the method to the cases of states and of events, there would be some of the same questions about hoping, fearing, noticing, and enjoying, but also some new questions about predicting, diagnosing, establishing the exact time and frequency of occurrence, and the like. For example, one might find oneself asking how often one means by 'Napoleon,' 'the man who lost at Waterloo,' and whether meaning this sometimes happens 'out of the blue' when one is not thinking about Napoleon, or whether it happens only when one mentions him, and if so whether it happens concurrently with, a little before, or even a little after mentioning him. And so on.

I do not propose to work out in detail the application of the method to the cases of states and events. Instead I will boldly claim that meaning is not an activity, a process, a state, or an event. This seems to me to make it highly unlikely that it is a mental or physical phenomenon at all; but there just might be some other category of phenomena, and so as not to rely too heavily on one argument I will advance some other considerations to support the same conclusion.

Something obviously called for is that I should go back over the arguments I suggested in the first place to the effect that meaning is something mental, because if what I am now contending is right, each one of those arguments must contain a mistake somewhere.

The first of them was the strong grammatical suggestion of two things, not one, contained in such remarks as 'I said it and I meant it,' or 'What I meant when I said that was ...' Has something gone wrong here? In the first place, only some grammatical constructions carry this suggestion: we say 'I said it and I meant it,' but not 'I said it and also meant it.' and certainly not 'I said it but omitted or forgot or failed to mean it'; and we say 'What I meant *when* I said that was ...,' but not 'What I meant *as* I said that was ...'

We do not, in short, express ourselves in all the ways you might expect if there were two things here; but still you might ask whether, grammar aside, it is not anyway the case that 'I meant it' or 'I meant such and such' reports something. I will try to show that it does not, by giving an alternative account of the conditions under which we say we mean something. This account will apply only to cases of 'meaning what we say,' and different accounts would have to be given for other uses; but from it you may at least see that we don't *always* have to take these expressions to be reporting something.

I am going to suggest that when we make a dire threat or an extraordinary claim and say that we mean it, our saying this is not a report of anything, but an action, which could be compared to a move in a game like chess, or to a promise: it puts us in a new position where it will be particularly embarrassing if we do not carry through on our threat or stand by our claim. There may be planning and deliberation prior to the move or the promise, but the move doesn't report the planning, nor does it report a decision, because one has moved or promised just as certainly if there is no deliberation or decision. Making a move or promising just are performing certain overt actions in certain circumstances.

An obvious objection here is that, although saying that we mean something does indeed have the social consequence of making it particularly embarrassing to retract, still either it is just because we have professed the existence of a certain mental state that those consequences follow, or at the very least it is because we have the mental state that we are prepared to take the action and risk the consequences.

The answer to this objection is, first, that the existence of a mental state has nothing whatever to do with these social consequences following. There is no imaginable reason why it should be more embarrassing to default on a threat if you felt tense about it when making it than if you made it quite unemotionally. This can be seen particularly clearly if you imagine a threat followed, not by 'I mean it,' but by a description of this or that mental state. If you try this with various possible descriptions, such as 'I feel tense,' 'I feel a sort of steely calm,' you will find that if they have any consequences, they are very different from those of saying 'I mean it.' In one case the response would be 'How interesting!' in another, 'How touching!,' but in none will it be anything like 'You wouldn't dare' or 'We'll believe it when we see it.' Nor would such descriptions quite answer the question 'Do you mean that?' Any answer to this question other than 'Yes' shows hesitancy, shows that one is not quite prepared to say that one means it.

It is simply the way the game is played, that default is more embarrassing if you have uttered the magic words 'I mean it' than if you have not done so. It is the words, not anything they stand for, which carry the consequence.

Nor is it true that it is only because we are in a certain mental state that we are prepared to make the move and risk the consequences. We do not, or anyway, I do not, first notice the mental state and then decide that in view of its existence and its properties it would be reasonable to go so far as to say that I mean it. Do we not, quite often at least, just say that we mean it? And is it not (quite often) the awareness of the new plunge we have thereby taken which stirs the emotions characteristic of meaning a dire threat, and also makes us resolve to be as good as our word? We are, I think, predisposed to think that we first mean it and then say we mean it; but in the kind of case I have just been suggesting, our saying that we mean it comes first, and from this other things result. I am not saying that it is always this way. The interesting thing is that this kind of case is both possible, and when you think of it, not uncommon. And what this shows is that we are not bound to the model according to which our saying things is secondary, and must report or express or describe something in us, which is the primary thing.

We have in the foregoing also provided a partial reply to the second reason given earlier for thinking that meaning this or that is something mental. That reason was that often there are mental states that occur at the time we say we mean something which are, the argument suggests, obviously what we are reporting or expressing in so saying. For example, if I feel defiant when I say something paradocical and mean it, my saying that I mean it will on this view be a way of saying that I feel defiant about it.

While we have already countered this argument in one way, there are other things against it. It is in the first place fairly obviously false that when I say I mean it, I mean that I feel defiant (or belligerent or calm or earnest or anything like that) about it. It is hard to be very clear as to what we do mean when we say that we mean it, but I suggest it is something like this: that in spite of any reasons for suspecting otherwise, what we have said can be taken at its face value – it means just what it appears to mean. And isn't this why we can't mean things whose meaning is not in itself fairly clear? I can't mean it when I say that the St Lawrence River is in love with the Bay of Fundy, because I haven't yet said anything which is meanable, which can be taken at its face value. It has no face value.

In case you are unconvinced by this account of what it means to say that you mean it, perhaps I might add that there are so many different mental states which might accompany our saying this, that if 'I mean it' were thought to report the existence of one of these, there would be no way of deciding which of them was being reported. When I make a dire threat I may do so in various moods: of exhilaration, steely calm, or intense anxiety. Which of these is reported when I say that I mean it?

To this question you might be inclined to reply that we at least know that one or other of a family of such states exist, and that is what is important. We don't always need to know which one it is.

The shortest of various possible answers to this objection is that the existence of one of these states is simply not what is important. Let us imagine a case in which a person says that he means something, but cross-examination satisfies us completely that nothing whatever happened except that he said it and said that he meant it. The absence of any member of the steely calm/intense anxiety family, if it is a family, would in itself give us no reason whatever for thinking that he didn't mean it. It is not *that* information about a person, but rather such things as its being entirely out of character, or there being reason to see the threat as a joke or a trick, or his taking no steps at all to follow through on it, which show whether he meant it or not.

Nor is it, when you think of it, anywhere near true that there is generally a mental state associated with our saying that we mean something. We fasten on a few cases where it is true, and quite overlook the great mass of them where it is quite obviously false. Here are some examples: (1) 'You

said you bought a car. Did you mean what we in France call "un auto"?'
'That's right.' 'And did you think that as you said the word "car"?' 'Of
course not.' (2) 'You said it is hot. Did you mean to say that it is humid?'
'Yes I did. It was careless of me.' 'And did you think "It is humid" as you
said "It is hot?" ' 'Of course not. I just see now that "humid" would have
been the better word.' (3) 'Did you mean what you said just now when
you said it is a nice day?' 'Of course I did. I nearly always mean what I say.'
'And how did you do it? Did you mentally mark "Correct" beside it as
you said it, or did you feel a glow of sincerity?' 'Of course not. I just said it.
What do you take me for?' Such examples could be multiplied endlessly,
and would show that we really hardly ever have an accompanying mental
state of any interest when we mean something.

There is, of course, the other type of case, for example where we make
terrible threats or extraordinary claims, which it is 'hard' to mean, or
where, having attached an unconventional meaning to an expression, we
are conscious of its oddity as we use it. But I think we unbalance our diet,
feeding only on these cases and coming back to them again and again. And
we do this, not because they are in themselves so convincing (I think I
have shown this is what I have just now been saying) but because we are
convinced that there must be something which 'to mean' stands for, and
since these are the only examples which are even plausible, we use them
to satisfy our hunger.[2]

Let me go on now to the third of the general considerations which I
suggested earlier tempt us to think that meaning is something mental: the
fact that it can be true or false that a person means something, and the conse-
quent hunch that it is true if an appropriate mental state exists, and false
otherwise.

Here again there is the empirical difficulty that although there are
exceptions, and we do fasten on these, still, as we have just now seen, in the
general run of cases where it is true that we mean something, there is
nothing that will at all nicely serve as the mental state that makes it true.

We are inclined, however, to be undaunted by this fact because, we
think, mental states are queer things, hard to pinpoint and describe. Isn't
it, one wants to say, only a problem of discerning and finding words to

2 Wittgenstein: 'A main cause of philosophical disease – a one-sided diet: one nourishes
   one's thinking with only one kind of example' (PI §593); and 'You think that after all
   you must be weaving a piece of cloth: because you are sitting at a loom – even if it is
   empty – and going through the motions of weaving (PI §414).

report these elusive phenomena? We shouldn't expect that things like this will be obvious, or easy.

We are to suppose then that there is something in us when we are trying to say what we mean that serves as our guide or source of information as to what to say. We examine it and notice things about it, and we are in possession of some sort of knowledge or principles by means of which we convert what we notice into literate sentences that say what we mean.

Is that how we tell in our own case whether we mean something? The question how we tell implies that there are some procedures that we follow; but if you remind yourself of what happens when the question is raised what we mean or whether we mean such and such, you will see that most often we follow no procedures. We don't call up states of mind and check whether they are adequately expressed in sentences or paragraphs, *or do anything else*: we just straight off say what we mean, or straight off answer 'yes,' 'no,' or 'not quite' to the question whether we mean such and such. Well, perhaps not always straight off, exactly: we often feel a momentary tension, furrow our brows, or pound our forehead, and then say something. But none of these things is what it is to mean something. They tend to be the same whatever we mean. By 'straight off' I mean 'without any prelude that is logically connected with what we end up saying.'

Still, there *are* cases where we do follow a sort of procedure. One very common way of telling whether such and such is what we mean is by whether, when we have said it or thought it, it strikes us as inept, unclear, incoherent, or untrue. If on any such grounds we reject a way of putting it as 'not being what we mean,' it is not that what we have considered saying fails to express what is in our minds, but that it strikes us, when we examine it, as not being up to certain standards. In cases like this there is no important difference between misgivings we may have about what someone else says, and those we have about what we find ourselves saying. And one could hardly say that our misgivings about other people's remarks are due to their failure to express our state of mind.

Another kind of case where we know that what we said we meant is not in fact what we meant occurs where for whatever reasons we have deliberately misled; but in these cases it is having decided to mislead, rather than perceiving a discrepancy between what we have said and our state of mind, that shows us that what we said is not what we meant.

In general what I would say about meaning is this: that in the normal case we just open our mouths and say something, and there is neither any question of whether we mean it nor any positive way of showing that we do. If the question does arise, it is not on the grounds of any positive correlation between what we said and what we had in mind to say that we decide we do mean it, but on the grounds of the absence of any conditions to the contrary: the absence of any decision to mislead, or the absence of any misgivings about the truth, coherence, clarity of what we said.

In an average case in which we have said something, it is not exactly that *of course* we mean it, but rather we do not right away know what to make of the question whether we meant it. We want to know why the other person asks: does he think there is something paradoxical, obviously false, or otherwise off-colour about what we said? We would best reply to the question whether we meant it, not by assuring him that we did, but by showing him that what we said seems to us in no way off-colour, and therefore we can not see how the question whether we meant it arises.

I have said that we both can and do say what we want to say directly, straight off, without any such procedures as describing a state of mind, or comparing what the words say with what we want, wish, mean, or intend to say. By this I do not mean that we always or generally speak without effort, that we do not hesitate, agonize, clap our heads, or furrow our brows, but only that whether or not we do this, there is still no process of deriving or constructing what we will say from mental or other materials. Since many people may have difficulty accepting this view, I will devote a little time to constructing and commenting on a contrary picture I think we are very much tempted to paint.

According to this picture there is such a thing as knowing what one wants to say before one finds the words in which to say it. This knowledge takes the form of a pre-verbal mental content, sometimes but not generally a mental image. It cannot generally be an image, because for most of the things one ever wants to say one could not suggest an appropriate image; being neither pictorial nor verbal, it comes out as something queer, indescribable. The struggle to 'say what we mean' is a struggle to encapsulate this queer thing in words. Words are here thought of as a kind of code into which mental contents are translated for purposes of communication and storage. The words in themselves are as dead and meaningless as a code generally is: they acquire meaning and are understood by inducing a mental state in the minds of people who read or hear them. And they are correctly understood, and the process of communication is successful, if the states of mind of the speaker and the hearer, or the writer and the reader, are more or less the same. According to this account, this is why we never quite know whether a person has understood: we only have external signs of his state of mind. If we could see into his mind we would know for sure.[3]

But now let us have another look at some of the parts of this picture; first, the idea that 'knowing what one wants to say' is having a (queer) mental content; second, the theory that saying it is encapsulating this content in words; and third, the belief that another person understands these words when they have produced in him a mental content reasonably like the one encapsulated.

What is it to know what one wants to say? In the first place, as I suggested earlier, in many cases it is entirely, and in all cases it is partly, a matter of wanting to say things of certain kinds. If we sit down to write a letter, we want it to be friendly, newsy, well-written, amusing; or if we set about making a point in a discussion we want it to be relevant, coherent, clear, true, and so on. Asking ourselves whether a sentence expresses what we want to say, therefore, is to this extent not a matter of determining whether it encapsulates something, but whether it satisfies these requirements. And the requirements are not generally something we have in our minds, but rather something which, as highly developed beings, are just part of the way we function. We react, as it were instinctively, to a clumsy phrase or an irrelevant point.

But what about the extent to which we do want to say something specific before we find a way of saying it? We think we know just what we want to say; but what can this knowledge consist of if it is not yet in words? One type of case might be that in which I want to make a point of a type that I understand very well and have made before in various contexts, but adjusted now to the nature of the discussion I am engaged in. The hard fact here, I suggest, is not that I have something not yet expressed in my mind, but just that being a person who understands the matter at hand, I will *of course* be able to explain it: give examples, answer questions,

3 In case you are inclined to smile at this picture, perhaps I might remind you how often it is found in philosophical literature, for example in Locke's *Essay Concerning Human Understanding*, Bk III, Ch. I, §2: 'Besides articulate sounds, therefore, it was farther necessary that [a man] should be able to use these sounds as signs of internal conceptions, and to make them stand as marks for the ideas within his own mind; whereby they might be made known to others, and the thoughts of men's minds be conveyed from one to another' (see also Ch. II, §§1, 2, 6). For a more recent example see the passages from Jerrold Katz, *The Philosophy of Language*, quoted in essay 7 below, pp. 148-9.

repair misunderstandings. If the thing to be explained is a point about the internal combustion engine, it may be that, to assure myself that I do understand it, I imagine pistons jigging up and down and thereby turning a crankshaft. But this image is not itself my understanding of that feature of the engine: rather it is because I understand that I am able to generate the image. And if I draw a picture on paper of pistons jigging up and down in the course of explaining the thing to the other person, it is not necessary to suppose that the picture is a copy of the image, because it is as easy or as hard to understand how I could just draw it directly as it is to understand how I could just imagine it directly. Or for that matter, how I could just directly describe it in words. I might first imagine and then describe it, but I might equally first describe and then imagine it. What lies behind any of these performances is just *me*, an intelligent human being with some explaining skills and some understanding of internal combustion engines.

This is one type of case knowing what one wants to say, and there are others, some of which I will consider a bit later. At this point, for want of a better place for it, I might mention the characteristic experience of 'having something on the tip of one's tongue.' One might be inclined to offer this as the experience reported by the expression, 'knowing what one wants to say.' But this is surely another case of settling for any old experience that will satisfy our demand that there must be something our words stand for. That it is not the experience of knowing what one wants to say can be seen very clearly, I think, from the fact that when we do find the words which were on the tip of our tongue, they are not an expression of the experience. We express the experience another way, namely by saying, for example, 'Oh dear, it's just on the tip of my tongue.' And if we don't find the words, then contrary to what you would expect, in spite of this experience we just do not know whether we knew what we wanted to say or not.

In view of these arguments, it may seem gratuitous to consider whether, in trying to say what we mean, we translate a mental content into words. But I might be wrong in urging that there is no mental content here, so let us suppose that there may be one, and ask whether it is by relating this to various proposed verbal formulations that we decide whether this or that is what we want to say.

To begin with a couple of obvious empirical points: first, what is suggested here is that something happens which is at least a bit like comparing a description of a woodpecker to a picture of one, to see whether the description mentions every element in the picture and relates the elements as they are related in the picture. But nothing at all like that actually happens when we try to find the right expression. What does happen is generally that we say straight off what we want to say (not, as I explained before, without struggle or head-clapping, but without what I called 'logical' preliminaries). And in those cases where there is anything more than this going on, it is not a matter of considering whether certain words express our state of mind, but whether they are clear, coherent, or nicely put.

But now the interesting question is this: all kinds of things might satisfy these requirements, that is, be relevant or well-formed, but still not be the particular thing one was anxious to say. How do we know that although this or that would be a good point, and nicely put, and so forth, it is still not the precise point we are groping for?

Here I think we may be bewitched by a certain model of how this is done, the model namely according to which we recognize, identify, choose things by comparing them to a prototype. This is certainly an obvious way of designing a machine to do this kind of work, and it is one of the ways people do it. We give someone a picture, and tell him to bring back something just like it, or we identify real woodpeckers by their similarity to the picture in our bird book. And we may therefore be inclined to think that this is the only way, and that there must be something like this going on when we identify this or that as just the point we wanted to make.

But when you think of it, this is far from being the only way in which we recognize or identify. For one thing, we identify from descriptions, and this without first translating the description into a picture and then comparing pictures. We see straight off whether the object satisfies the description. And if we do sometimes entertain images as an aid in this business, we often find that we have entertained the wrong ones, and that the object satisfies the description better than the images do.

More interestingly though, we identify or recognize without any paraphernalia. I meet a friend on the street and I know him: do I compare him with an image and find that they tally, or do I remember certain propositions about him and find that they are true of this man? No, nothing whatever goes on. I just know immediately who he is.

However this can be explained, and even if it can't be explained, it is certainly the way it is. And I suggest that it is the way it is also with knowing whether this or that is what we wanted to say. We simply accept some things, and reject others. You may feel like objecting that in that case we don't after all know whether this or that is what we wanted to say. And I don't know whether to agree or not. We generally do not have any doubt, and this is part of what it is to know, but there is normally not a scrap of evidence we could adduce.

Let me go on now to the third feature of this picture of the process of communication which I set out to discuss: the view that words produce a mental content in the person who reads or hears them, and that he understands if the content produced is the same as that in the mind of the author of the words. I suppose the main reason we are attracted to this view is that we think the word 'understand' must refer to something, something we do or that happens in us, and we can think of a lot of cases in which words do produce something in us, mostly cases of hearing or reading descriptions, when we imagine the scene or the object described. We want to say that the picture we imagine is the understanding, and that we have understood if we have the right picture.

It is very much less easy to suggest an appropriate mental state when it is anything other than a description we understand, so here again the diet may be unbalanced. But for brevity's sake let us concentrate on understanding descriptions, this being the most plausible instance.

We tend to convince ourselves that understanding is imagining what is described, by again and again thinking of a description and then imagining something. But I wonder how often we do the imagining in the ordinary course of events, when we read a description in a novel or a newspaper? We understand these descriptions, and hence no doubt we could imagine (or draw or recognize) the scene: but how often do we actually do such imaginings? If I describe to you a gracious room with tall windows looking out onto a secluded garden, you will perhaps now imagine it. But are you sure you would do this whenever you encounter such a description? And do you not understand it on these occasions when you do not imagine it?[4]

Then to what does the word 'understand' refer? The answer, I believe, is that it refers to nothing whatever – not a mental state, not a description,

---

4 Wittgenstein: 'We fail to get away from the idea that using a sentence involves imagining something for every word .... It is as if one were to believe that a written order for a cow ... had to be accompanied by an image of a cow, if the order was not to lose its meaning' (PI §449).

not a pattern of behaviour, nor anything else. The word 'understand' has various uses, none of which is of the kinds here contemplated. One such use is to explain the *sort* of difficulty one has with what someone has said, and to apply for a particular sort of further explanation. If I say 'I didn't understand that last sentence,' I am not saying that I query whether it is true, or that I am offended by it, but that I am uncertain what it means, and that I would like the other person to try saying it another way. But 'I do not know what it means' does not say that instead of a supposed normal state of affairs in which my hearing it is soon followed by my interpreting it to myself there is here an unhappy case in which the interpretation fails to ensue; but rather that the normal state of affairs in which one has no difficulty with what people say – in which their words themselves, as one might put it, express their meaning – does not in this case obtain. The normal easy play of conversational interaction has foundered, and in my distress I say I don't understand.

It is not upon satisfying ourselves that certain things are not true that we say that we do not understand; we just look distressed and say it. Do we say it *because* we feel distressed? It is more likely that we feel distressed because we do not understand than the other way around.

Negative and hypothetical uses of 'understand' are much the most common. We say 'I don't understand ...' and 'If I understood you rightly ...,' but we are a good deal more hesitant about saying that we do understand. This may confirm the above suggestion that the cases in which we understand are just the normal routine cases of linguistic intercourse between accomplished language-users. There is in these cases nothing peculiar that goes on, which might settle whether we understand. 'I understand' is, as Wittgenstein put it [PI §323; see also §180] a glad start, a cry of joy or relief. We let out this cry when some blockage in the normal conversational flow has been removed.

In saying that we understand, are we nevertheless saying that we have hit upon something that makes everything fit again, something in the light of which things once more make sense? Well, we perhaps sometimes *have* hit upon something, formulated it to ourselves, and considered how well it makes things fit; but most often nothing as *orderly* as that will have happened. If, with a sudden sense of relief and excitement I say 'I understand! What you meant was not that ... but that ...,' the words following 'what you meant was' will be a statement of what I have hit upon; but must I have thought that part of this utterance before I said it? No: the

expression of my new insight *could have taken form as I said it*. If you
wonder how that is possible, the answer is easy enough: I am acquainted
with distinctions of that type. I have seen them made in other connections,
and have made them myself. I may suddenly have realized that there could
be an application of this type of distinction to the case at hand; but that
realization itself need not have been spelled out in my mind. I may simply
have found myself ready to make this application, and have known just
what I was ready to do only when I had done it.

This is all I will say about this curious picture of the process of communication, or about the more general question as to whether meaning
things is a mental phenomenon. I cannot claim to have shown anything
conclusively about these matters, but at most only to have mapped out an
approach to them, and shown how that approach works for a few, and I
hope not the least difficult, aspects of the question. Because of the multi-
farious character of these questions, I believe one can in this manner only
provide some equipment with which a person may shift for himself. This
being so, it may be useful if in conclusion I review some of the tactics I
have employed.

As well as the way described earlier of determining whether something
is an action, a process, and the like, it will be noticed that a useful device
was that of finding a translation of sentences in which expressions like
'I mean,' 'I meant it,' 'I meant to say' appear, having the same sense as the
original sentence, but containing neither one of these expressions nor any
other which, like them, consists of a personal pronoun and a so-called
mental conduct verb: neither, for example, 'I meant to say' nor 'I intended
to say,' 'I wanted to say,' 'I was trying to say.' In any case in which this
can be done, it will remove the temptation to think, on grammatical
grounds, that any user of these expressions is reporting something about
himself. It will be seen that I have done this on several occasions. I have
suggested, for example, that sentences containing the expression 'I mean'
can be translated to read 'What follows is a further explanation of what
has just been said'; and that sentences containing the expression 'I meant
to say' can be translated 'It would have been better to have said.'

There are some similarities between this tactic and a move which con-
sisted of (a) substituting a description of a supposed mental state of meaning
something for the words 'I meant it,' and then (b) bringing out the difference of sense thus generated by showing the different roles the two
utterances would play in conversation. (To a dire threat followed by 'I

mean it,' appropriate reactions would be 'I don't believe you' or 'You had better reconsider'; while to 'I feel tense about it,' appropriate reactions might be 'How interesting,' or 'Yes, but do you mean it?')

A different sort of move has been that of 'balancing the diet' by assembling reminders of how very many of the circumstances in which we use these expressions are entirely unlike what we are inclined to imagine. We dwell on a few cases which seem to (but do not) fit the picture we want, and blithely overlook the great mass of cases which very clearly do not fit it. For example, when reflecting on what it is to say something and mean it, we dwell on cases of things which are difficult to mean, like dire threats and paradoxical claims, just because here there is generally a mental state in the neighbourhood, namely the feelings of defiance or resolution which we experience – and we overlook the great majority of cases of our saying that we mean something, in which nothing in our consciousness can be found which will pass as what it is to mean it.

Related to this is the method of *constructing* cases in which there is no mental content, or none that we could call what it is to mean, and putting it to competent users of the English language, that they would be perfectly happy to have the verb 'to mean' used in these contexts too. The reason for mentioning this as a separate device is that it can be used specifically against those cases which I suggested we fix on and feed on, the cases in which there does seem to be a plausible candidate for the mental state of meaning something. It is not enough to say that of course there are these cases, but then remember there are others: one must also come to grips with the beguiling cases themselves. One way of doing this is to describe a case of the same kind, but without the accompanying mental state, and show that a person can be said to mean it here too. The purpose of so doing is to make it clear that even where the mental state exists, it is not essential to 'meaning something.' This is why I described the case of a dire threat in which nothing whatever happened except that the author of the threat said that he meant it; and why I urged we would have to deny in these circumstances that he meant it.

A quite different sort of measure was that of examining the explanatory models which, if we think they are the only possible explanation, contribute to persuading us of the existence of certain mental states. If we think that the only way of identifying is by comparing, then we will be much inclined to suppose that when we identify a form of words as 'what we meant,' we must have compared it and found it to tally somehow with a pre-verbal mental state of wishing to say something. I have taken two sorts of moves against such models: first that of showing that the beguiling model is not the only possible explanation of how we could do such things as recognize; and secondly that of suggesting that even if we cannot explain how we do it, there most certainly are cases where at any event we do not do it in the way the model requires – the case of recognizing a friend, for example.

It may sound as if, by artful operation of machinery such as I have been describing, one should be able reliably to show that for example 'meaning it' is not a mental something or other. Unhappily this is not true. One tries in these ways to be open and rational about it; but it sometimes seems as if success in these matters is more like causing a loss of religious faith than like a triumph of fair argument. People are not driven rationally to the conclusion that God does not exist, they just find one day that they no longer believe it. And then they may be unsure even as to what to say it is, that they disbelieve. The conclusion of sceptical arguments no longer seems stateable, and therefore the arguments themselves seem unreal.

Again, arguing about these matters is like the treatment of a disease: one carefully applies the best medicine one can find, but for a long time the disease resists the treatment. One day, however, the patient recovers; and then one is left wondering whether possibly the recovery was spontaneous.

This curious state of affairs may be what Wittgenstein had in mind when he said[5] that his aim in philosophy is to show the fly the way out of the fly-bottle. There is a way out, and you really are indicating it when you stick a pencil in the neck or darken the bottom or tap everywhere except at the exit. But the fly is oblivious to such advice, and buzzes about the same way whatever you do. And if he ever does drift free, it will not likely be due to your having shown him how.

5   PI §309

Personal
identity

I

When a man acquires a new heart, it does not occur to us to ask whether he may thereby have become a different person, for example, whether he is still Philip Blaiberg. Most of our inner organs, as well as our arms, legs, and so on we regard as being what we might call 'identity-neutral,' that is to say we think that one of them could be replaced by another without a different person resulting. Anything that will do the same job, we might theorize, can be substituted without a change of identity; and this might apply even to parts of the brain. If there were an isolable segment of the brain that was responsible for mathematical calculations, and it could be replaced, even by an electronic calculator, the survivor of such an operation would be regarded as the same person, who had simply had some of his (non-essential) support equipment changed.

A person may, it is true, change quite fundamentally as a result of some transplant operations, and we may then say 'He is quite a different person now'; but by this we would just mean that he is more cheerful, more emotional, or perhaps less cautious. We would not mean that he is no longer the man who married Mary Smith, or no longer that blonde chap who sat in the front row in Grade VII.

There are, however, questions as to how far the principle that bodily parts are identity-neutral extends. We will allow different hearts and even mechanical hearts, and different limbs and even mechanical limbs; but changes are imaginable that would leave us in doubt. If an old man's head were transplanted onto a young man's body (not to mention a young *woman's* body), or if a human head were successfully affixed to a mechanical body, then although some people might still stoutly affirm that it was the same person, we are a lot less certain. Disagreement becomes possible. We might disagree similarly about other bodily changes: one which gave a person a new face, for example. For there is a difference between plastic surgery, in which the face is *altered*, and an operation which gives a man *someone else's* face.

Although disagreement is possible in such cases, there is, nevertheless, a very strong inclination to say that the survivor of the operation is the same person. But consider this kind of case: one man is suffering from a grave condition of the brain, but the rest of his body is in pretty fair shape, and the doctor recommends a brain transplant. Another man's brain is hale and hearty, but the rest of his body is in dreadful condition, perhaps as the result of an automobile accident, and the doctor recommends a body transplant. Suppose that these two men enter the operating room, one needing a new brain and the other a new body; the good brain is successfully transplanted to the good body, and the defective body and brain are abandoned. Call the pre-operative person with the good brain 'Peter' and the pre-operative person with the good body 'Paul': are we to say that after the operation it is Peter that has a new body, and hence that Paul has died; or that it is Paul that has a new brain, and hence that Peter has died?

There are enough lifeless parts to make one human corpse, and the world's population has been reduced by one; but because the parts are not the parts of one man, it is peculiarly hard to say *whose* departure has so reduced the population. And it is not unclear the way it is unclear who has died when, for example, the victim of a highway accident cannot be identified. In that case, due to lack of identification, we may not know who it is, but we have no doubt that some identifiable person has perished. In the double transplant case there is no missing evidence and we know very well who went into the operating room. We just cannot decide who came out.

The problem as to the identity of the person who survived the operation, and the correlative problem as to who should be regarded as having died, would be important to a number of different parties. It would be of obvious importance to the wives and families of Peter and Paul; it would be important to life insurance companies and succession duty offices; it would be important to courts in deciding as to the marital status of the survivor of the operation, and the disposition of the property of the pre-operative persons; and to persons with whom Peter or Paul may have had contracts to write a book, or to pose for photographs; and to the police, in deciding as to the criminal responsibility of the survivor for the pre-operative acts of Peter or Paul.

The question of the identity of a living being, or of a dead one, is normally decisive in all such contexts: if this is the man who did the embezzling, then criminal responsibility attaches; if this is the corpse of the person with whom I made a contract, then the contract is at an end; if this man is Paul, then he has no right to Peter's property, and a divorce will be in order if he wishes to leave Paul's wife and live with Peter's wife. Because generally all we need to know to settle such questions is who a person is, we will be naturally disposed, in the hope of coping with the legal, moral. and emotional problems arising out of the brain transplant, to press the questions who the survivor of the operation is, and who it is that has died.

These questions might appear to arise only if two persons who are still legally alive enter the operating room, while only one living being emerges. For if the condition of Peter, the man with the ruined body, were allowed to deteriorate until he was by some definition legally dead before the transplant occurred, then we might think that his wife would be a widow, she would collect his life insurance, the books would be closed on any crimes he had committed or on any contractual obligations that could be discharged only by him personally (rather than by his estate).

Still it will easily be seen that such a solution is unsatisfactory. Delaying the transplant until one of the donors is in some sense dead does not in an essential way change the complexion of the situation. If such operations were routinely possible but could best be done with living donors, there could only be a legalistic reason for delaying the operation until one of the ailing parties 'died.' This can be seen particularly clearly if we imagine a case in which Peter and Paul both have a very short life expectancy, and the doctors simply wait until one of them, it does not matter which, dies, before performing the operation. Then on the one hand regardless of who dies first, the post-operative situation will be the same: we will have Peter's brain in Paul's body, Peter's wife deprived of a well-loved face and Paul's of a well-loved personality, but on the other, which wife is the widow, who collects the insurance, what contractual commitments the survivor has, whether he will be an adulterer if he goes home with Peter's wife, and so on will all depend on whether Peter or Paul happened to be the first to 'die.'

It would be absurd that the survivor should be Paul if Peter happened to die first, but that without being in any material way different, he should be Peter if Paul happened to die first.

If the problems cannot be solved in this legalistic way, how can they be settled? Three different sorts of things might in fact *happen*, but whether in happening they would contribute to showing who the survivor was, we shall have to discuss.

Here and throughout this essay we will assume that the survivor of a brain transplant would have substantially the knowledge, memories, tastes, ties of affection of the person whose brain he has. This *might* prove to be altogether false, or true only with qualifications; but we do not need to worry about that. The possibility of successful brain transplants is in itself utterly remote; and we are not engaged in this exercise because we want to be ready with an approach to problems of identity when brain transplants start happening. We are using the logical possibility of brain transplants to cast some light on problems of personal identity in general, and since for this purpose we do not confine ourselves to practical possibilities, we are free to make whatever supposition may serve some philosophical purpose.

Now the first *kind* of thing that might happen is that various of the parties intimately involved might regard the survivor of the operation as being Peter, or as being Paul. Peter's wife, for example, might accept the new person as being Peter in spite of the radical physical difference, with no greater sense of strangeness than if his face had been disfigured in an accident and repaired by plastic surgery in such a way as to leave him looking quite like a normal human being but quite unlike Peter. But if she had previously known Paul, she might find it simply impossible to see Peter in Paul's features. She might find that she could *accept* the new person as a husband, but that it is rather more like being remarried than continuing the same marriage.

Paul's wife, on the other hand, might find it impossible to look on that familiar face and believe that it is not Paul. She will, of course, be disconcerted by the personality differences between the man she now knows and her husband, but she would not necessarily find this more distracting than if her husband had developed some acute mental affliction that affected his memory and his personality. She might find herself as unable to doubt that it was Paul who had changed in this distressing way as we generally do when someone undergoes a radical personality change.

Being accepted by this or that person, however, or indeed by any number of persons, as being Peter or Paul, will not decide who the new person is. We could not say that if Peter's wife, who knows him so well, accepts the new person as being Peter, then this is surely at least very strong evidence that it *is* Peter. It is not as if the survivor of the operation might be pretending to be Peter, and while most of us would not know enough about Peter's personality to be able to detect a fraud, his wife would. For none of us has any doubt as to the facts in this case. It is part of the *problem* that the survivor has Peter's personality, memories, and so on and therefore a wife's testimony that he does indeed have her husband's personality can not logically contribute to solving the problem.

A second kind of thing that might happen is that if brain transplants became a common and routine practice, we would no doubt work out legal principles governing such questions as which, if either, marriage remained in effect after a transplant, which wife could collect on her husband's life insurance, what contractual obligations would remain in effect and whether criminal responsibility would attach to the survivor for the acts of either of his predecessors. But decisions about such questions would not simply follow in the train of a decision as to who the survivor really is, nor would they show who he is. One can imagine some such legal principle as the following being adopted as a reasonable solution to certain contractual problems [for convenience let us, without prejudice to his real identity, call the survivor of the operation 'Ezra,' and let us refer to a party with whom either Peter or Paul had contractual arrangements involving their rendering him personal services as 'a contractor']: 'A contract involving the physical appearance or physical strength or dexterity of Paul shall be binding on Ezra, but voidable by a contractor upon proof that the transplant incapacitates Ezra for the performance of the services called for in the contract. A contract involving the knowledge or intelligence of Paul shall not be binding on Ezra. A contract involving the knowledge or intelligence of Peter shall be binding on Ezra, but voidable by a contractor on proof etc.'

In this we see how it might reasonably be laid down that for certain purposes it would be as if Ezra were Paul, while for other purposes it is as if he were Peter. But if in a given case Ezra were deemed to be as it were

Paul, that would show neither that he really was Paul, nor even that for legal purposes he should be treated *across the board* as if he were Paul. For Paul might have had an agreement to pose regularly for a photographer, while Peter had a contract with a publisher to write a book. Under the former of these contracts Ezra would on the above principle be treated as if he were Paul, and under the latter the same principle would prescribe treating him as if he were Peter.

A third possible solution, apparently more decisive than either who Ezra is accepted by family and friends as being, or who he is deemed by law courts to be, is who he regards himself as being. We might expect that when he came out of the anaesthetic Ezra would surely take himself to be either Peter or Paul, and by all odds most likely the former. He would, for example, if he asked the nurse when his wife would be in, be thinking of Peter's wife, and would be likely to be cross if everyone including Paul's wife behaved as if *she* were his spouse. He would worry about Peter's mortgage payments and want to get back to Peter's job. Having known beforehand how the operation would change him, he would not likely be shaken in regarding himself as Peter by looking in a mirror, or by the fact that his friends did not recognize him, or because other people greeted him as Paul.

Certainly in these circumstances anyone not intimately affected by the question who he was would let it pass that he was Peter. We would see no point in persuading him that he was Paul, or that he was neither Peter nor Paul, but someone else. But this would surely be less because we took his not doubting he was Peter to show that he was in fact Peter, than either because, not ourselves knowing *who* he was, we would have no reason to press for any other view, or because we would see no point in upsetting him in this way. If Paul's wife persisted in regarding Ezra as Paul, we would not, just because he regarded himself as Peter, treat it as out of the question that she could be right. We would not attribute to her the elementary blunder of supposing that he did not regard himself as Peter when he most obviously did; or ask her whether she thought he was just pretending that he was Peter; or whether, if she did not think he was pretending, she at least thought he was in some condition akin to schizophrenia, such that although for the time being he genuinely had Peter's attitudes, memories, abilities, he could conceivably some day switch over to having Paul's.

Paul's wife need not believe any of these things. She could say: 'I am not believing anything false or improbable in believing he is Paul. Anything that any cool observer believes about this man I am prepared to believe; but if you say he is Peter in spite of having changed in certain fundamental respects, I don't see why I can't equally say he is Paul in spite of fundamental changes. Paul is no longer fond of me, no longer says the funny things he used to say, and no longer remembers the holiday we had last summer. From my point of view he has changed most regrettably; but I still say it is Paul who has changed in these ways.' Such a declaration ought not to convince anyone that it is indeed Paul: but it does show that it is not absurd or unintelligible to believe Ezra to be Paul.

If we now imagine Ezra (still supposing him to have no doubt that he is Peter) to listen open-mindedly to what she has to say, not only ought we to say that he no more than anyone else should be convinced by what she says, but we can scarcely conceive what it would be like for him to be so convinced. Still, what she had to say might very well make it much less clear to him *who* he was; and we can imagine him saying on due considera- tion, 'I am no longer prepared to affirm at all confidently that I am either Peter or Paul. All I know is that with the memories, emotional attachments, and tastes that I have, Peter's is the life I can accept with the least disruption and adjustment. It is not that there are *no* disruptions involved in my being as it were Peter. My children treat me as a stranger, and find it weird that I should make references to things I have said and that we have done together when to their minds these are in fact things that Peter has said and that he and they have done together; and even my wife is visibly shaken by my appearance often. It is distressing too to be taken by Paul's friends to be Paul, especially when I do not know whether to say that I am not Paul, but only have his body, or that I am Paul, only I will have to be forgiven if I don't remember them, since I now have Peter's brain. I can see how I could, by pleading for this kind of forgiveness, become as it were Paul, and in some ways I would find that not unnatural. But on the whole, given that I must accept Peter's life or Paul's, I am overwhelmingly inclined to accept Peter's. The crucial thing for me is that, whatever *they* feel, I have no difficulty in regarding Peter's family and friends as my own, and the same with Peter's books, records, paintings, and the like.'

Here we see Ezra no longer certain who he is, quite able to contemplate the idea of being Paul, but all things considered deciding, not that he is Peter, but to accept Peter's life. The question who he really is has been eliminated; and only the question who it is best to be remains. Moreover, this latter question would by no means always be answered in such a way that the survivor accepted the role of the person whose brain survived. We can see how, in a situation in which, for example, neither Peter nor Paul had been married, Peter had had few friends, and Paul's friends had prized him less because of his conversation and his tastes than because he was a fellow human being who happened to be part of their social milieu, it might seem best to Ezra to accept Paul's life. And more clearly we can see how, if neither Peter nor Paul had had any very close personal attachments, and if they had lived in a country where identity cards and income tax and succession duty offices did not loom large, the question who he was might scarcely have arisen for Ezra, and he might simply have proceeded to live his life as it came to hand.

A question similar to that of who the survivor regards himself as being is that of who, prior to the operation, might regard himself as being about to die; of what manner of farewell would be suitable in the case of Peter, and of Paul. Since the operation would be extremely risky, they both, of course, would be likely to make a will, and say many ultimate things to their family and friends. But for Peter, one might suppose, it would be reasonable to part with mutual expressions of hope for the future; whereas for Paul and his family the parting would be tantamount to a death scene, since however successful the operation may be, for Paul it will be the end of whatever knowledge, intellectual skill, wit, tastes, memories, idiosyncrasies, and ties of affection contribute so largely to making him what he is. The survivor of the operation will not remember Paul's wife or his children or any of the colourful or dull events of Paul's life; and not because he has forgotten them: he never knew them. Paul's farewells should be final. This may suggest very strongly that Paul should be regarded as dying, and that the survivor should be regarded as being Peter.

Two considerations, however, will make this inference much less convincing. The first is that, of course, for Paul more than for Peter, it will be the end of a great many of those things that we are likely to prize most about him. There would be a very great loss from the set of things that go to make up the pre-operative Paul as we know him; but it is no clearer that the loss is equivalent to Paul's ceasing to exist than it is in the case in which a person suffers a radical mental disorder.

Secondly, the transplant case is not importantly different from an imaginable case in which a person is suffering from some acute affliction of the brain, and the doctor tells him that he will soon die of this disease unless he takes an injection which will almost certainly cure the affliction, but which will result in the complete obliteration of everything he has learned since childhood, so that he will have to start all over again learning to speak, read, swim, ride a bicycle, and will remember absolutely nothing about his life prior to the injection. In such a case, whether to take the injection or not is not a choice between one death and another. It is not a matter of indifference whether to do it or not. Grim as the decision might be, a man and his family could regard it as a desirable step, permitting him to survive. If he survives, he will of course be very different, but it will be the same man who is different: it will be James Moriarty who prior to the injection was a marvellous raconteur, and who is now slowly learning to talk.

We have now canvassed a number of possible ways in which we might hope to decide who Ezra is, and found them all defective. In asking who he is, we tend to assume that Ezra is in fact either Peter or Paul: we may not *know* which of the two he is, but in the same way that although we may not be able to discover the identity of an accident victim, he does have an identity, so also the survivor of the brain transplant has an identity, if only we could discover it.

Who he is may seem to us to be a question of what being a certain person consists of: does it consist of having certain memories, attitudes, tastes, ties of affection, or does it consist of having a certain appearance, being short and wiry with expressive hands and an aristocratic if ugly nose?[1]

Put *this* way, and if we *had* to make this choice, most of us would no doubt be inclined towards the former alternative. But it is not so clear, when you think of it, whether we are thereby expressing an opinion as to what *it is* to be Peter or Paul, or simply indicating which of people's attributes we care most about. Another day we might be asked whether to be a certain person is to have certain personality traits and a certain appearance (thus including the question of appearance on one side of the choice), or whether it is to have a certain history. Here we might at first incline to the former alternative, but not if we were in search of a person missing for twenty years, whose personality and appearance could be assumed to have changed, perhaps beyond recognition.

---

1 Here it may be noted that it is possible to be a good deal more cavalier about dismissing 'the body' as essential to personal identity when it is referred to in that very general way. We can be much less confident in our dismissal when it is a question of expressive hands, aristocratic noses, or any other physical characteristics of people that we prize, or that contribute to their individuality.

It is thus partly only the accident of being obliged to make this particular choice that leads us to say that personality is essential and appearance is not; but even within the arbitrary limits of this choice it is not so clear that every reasonable person would decide in favour of personality. The police would not, for example, but would initially anyway be very moved by pictures, scars, and fingerprints. And which criterion some moving picture studios would use would depend on whether their contract with a man was for services as a scriptwriter or as an actor.

Nor is it only people with special interests who would be swayed here by considerations of appearance. If it was not Peter and Paul who went into the operating room, but Peter and Jane, then *any* of us would be likely to say that the bodily difference was essential.

It is not always a choice between appearance and personality criteria. We are prepared to allow that a person may change beyond recognition in both these ways, and still be the same person. On the basis of fragments of decisive evidence a mother may accept as her long lost son someone she cannot recognize by his appearance or personality. She may then worry about his fortunes, chide him about his smoking, miss him when he goes away, and tend his grave if he dies. Her acceptance may be complete; she need not be *resolutely* worrying about his fortunes or *resolutely* ignoring the strangeness of his appearance and mannerisms. She may be concerned about him not because she has decided to treat him as her son, but because she has no doubt that he *is* her son. She, however, not only believes this, but she has or may have sufficient reason to believe it. She may, for example, through both his testimony and that of other people, be able to trace his history back to the day when they were separated. There may, of course, be some mistake or some fraud involved here, but there is no question of anything queer having happened along the way. He is either her son or he is someone else, and there is nothing in between; and if the testimony she accepts is reliable, then he really is her son. We may have doubts as to whether his history is continuous with that of her son, but we have no doubt that *if* it is, then he is indeed her son.

Any doubts as to who he is are a matter of missing evidence: lost identity cards, blurred fingerprints, incomplete observations or character or personality, witnesses who cannot be located or whose testimony is suspect; but fill in that evidence and we will have no doubt whatever as to who a person is.

Compare this now with the case of Ezra. It is not *information* about him that we are lacking or that we distrust. We either have, or can readily get, all the information about him that we could wish, and still we remain confused as to who he is. This suggests that the trouble lies in our trying to treat the question about Ezra as an ordinary question of identity, a question, that is, in asking which we assume that Ezra is some known person, and the problem is one of establishing which person he is by assembling evidence: fingerprints, scars, birthmarks, testimony of continuous acquaintance, intelligence, personality, knowledge, memories. But since we know all such things about Ezra and still do not know who he is, we surely have to abandon that frame of reference. And since it is the frame of reference that gives sense to the question, in abandoning it we will be abandoning the question who Ezra is. Instead of assuming that he is some known person, our approach will now have to be that he is neither Peter nor not Peter, and neither Paul nor not Paul, but is just what we know him to be: a person having Peter's brain and Paul's body, and whatever abilities, personal attachments, and hang-ups result from that union. The complete Peter is no longer with us, nor is the complete Paul, but there is a new person with some of Peter's characteristics and some of Paul's, and in all likelihood some all his own. Either we cannot meaningfully ask the question who he is or there is a short answer to it: he is Ezra. Ezra is a person who has the peculiarity and misfortune of having two histories, each involving the rich array of rights, obligations, and ties of affection and acquaintanceship that surround a human being. To the extent that the relationships from either history survive the transplant operation, they will be difficult to reconcile, and no way in which they ultimately shake down is likely to be entirely satisfactory; but they are to be settled if at all by a shaking down amongst the parties affected, and not by any philosophical wizardry purporting to show that Ezra is really Peter or really Paul.

II

What is philosophically interesting about our question is what it may show about the concept of a person. What we may have been thinking in asking whether Ezra is Peter or Paul is that there is some constituent of a person which is there and remains the same as he grows older and changes, and it is this constituent we are identifying, or at least supposing the continued presence of, when we say, for example, that this is Mrs Carmichael's long lost son. According to many religions this constituent is the soul, and it unites with the body some time around birth, leaves it around the time of death, and could conceivably, even if it rarely does, shift from one person to another. The question who a person is is the question which soul is involved with a given body. A body may not need any particular soul to animate it, but every body needs some soul; and since souls are discrete entities, that is, are needed to animate bodies not as electricity is needed to make a motor go but as a pilot is needed to make a plane fly, it is always a reasonable, if sometimes an impossibly difficult question, *which* soul is now animating any particular body. This, according to this view, is what we are really asking when we ask whether the survivor of the brain transplant is Peter or Paul; and it is because of this conception, or of conceptions of this type, of what it is to be a certain person, that we tend to be so sure *a priori* that the survivor is some definite person: possibly Charles or Herbert, but by all odds most likely either Peter or Paul.

Our discussion so far makes it ever so much less clear that this is a viable picture of what it is to be a certain person; but lest the foregoing arguments be thought simply to *assume* the falsity of the picture, some further points of a more general kind may be added.

It is part of the picture of a soul in a body that the soul remains in some sense the same, not only while the body changes, but while the personality changes. If there is the same soul, there is the same person, regardless of the extent of bodily or personality changes; and if the soul is different, then even if the body or the personality is the same, we have a different person.

There are, however, two possible accounts of what it is for the soul to be the same or different: either it may change in whatever ways souls may change and still remain the same soul; or any change in it makes it a different soul. Both accounts lead us into difficulties.

On neither account is it easy to see how souls can have such identifying characteristics of people as aristocratic noses or expressive hands; and it is only somewhat easier to see how souls can have such characteristics as knowledge, memories, wit, or taste. (When could we say that John does not remember this, but his soul does, or that John knows this, but his soul does not?) But we will have to work, however uneasily, with the supposition that souls may have the latter sort of characteristics.

If we treat it as a possibility that a soul may remain the same soul while its characteristics change, then it will also be possible that a soul's characteristics should at least normally be those of a person – for example, that when my tastes change, normally so do those of my soul. We will then be able to make some sense of the idea of identifying a soul by its characteristics –

for example, of being able to tell with whom one is reunited after death. However, on this basis we are foreclosed from just the criterion of identity we were trying to offer in the case in which a person's knowledge, memories, tastes change radically through time. We wanted to say that he is the same person if he has the same soul; but if the soul changes as the person changes, it turns out that we have only swapped our problem for one of exactly the same form, but less tractable: what makes a soul the same soul when it has changed radically? There would be no progress in saying that it is the same soul if it has the same super-soul.

On the other hand, if we say that any change in a soul makes it a different soul, then since it is a datum of the problem that people can change very much and still be the same person, that is (on this theory), have the same soul, it will have to be the case that there is at most a loose relation between characteristics of people and characteristics of their souls. If Abigail's characteristics change drastically over a period of time, but from her being the same person we infer that her soul is unchanged, then at the very most her soul's characteristics could coincide with hers only at some one point of time.

We would perhaps have two choices here: to say that the characteristics of souls are of the same type as those of people, but cannot be straightforwardly inferred from those of people; or to say that they are of a different type, and not detectable in any of the familiar ways in which we come to know a person.

On the latter basis we would find ourselves having to discover or invent the characteristics. They would be properties heretofore unknown to the human race; and prior to our success in this venture only God would have known who anyone really was. Moreover, it is difficult to see how, without ourselves knowing in some other way how to decide questions of identity, we would have any case for saying that any new characteristics we did discover were identifying characteristics, rather than just heretofore undetected properties of people, which could change without entailing a change of identity. If we discovered that there was some characteristic that was in fact unalterable throughout a person's lifetime – having a beating heart might be close enough – there would be two difficulties: one, of identifying an instance of this property as characterizing a particular individual; and the other of showing that an alteration or discontinuation of this property makes someone a different person. For we may cease to be a person when our heart stops beating, but we do not become a different person.

The possibility that souls have characteristics of the same type as persons but that the two are not straightforwardly related may therefore seem more promising. Here our problem would be to invent or discover, not new characteristics, but new auspices upon which to attribute them: we have not heretofore routinely attributed characteristics to souls. Here again, if we found that we could seriously make such attributions, there would be a problem as to why we should regard any characteristic attributed as an identifying characteristic rather than as a heretofore undiscovered property. If it is possible for a person to hate his mother although he does not think he does, this is *like* the possibility we are now contemplating that persons and souls should have characteristics of the same type, not straightforwardly related. Hence we *might* in such a case attribute the hatred to the soul; but we would still have the problems, first of whether this instantiation of mother-hatred identifies this person (that is, of whether the mother-hatred would not itself have to have peculiar characteristics, which themselves might change, in order to be precisely Harry's mother-hatred); and second of whether Harry would cease to be Harry if his soul stopped hating his mother.

The belief that the soul is the continuing element is only one, and perhaps the most intelligible, expression of the view that what we are identifying or supposing the presence of when we say that someone is the same person is something or other that has remained the same, or the same x, while other things have (or have not) changed. When we are puzzled as to whether Ezra is Peter or Paul, even if we do not believe in souls, we are perhaps hoping that by careful thought or close investigation we may discover some real feature of a person that is the same when he is the same person and different when he is a different person. Exactly the same difficulties, however, apply to any such secularized version of the soul theory of identity: whatever the something-or-other is, on the one hand it must have individuating characteristics, otherwise its presence would not show who a person is (but at most only that he is a human being), while on the other hand it will not have any of the characteristics by which we do identify people, since it is agreed that all such characteristics can change without our being bound to say that we have on our hands a different person.

If one were therefore to say that what is needed is a radical refinement of our crude and haphazard technique of identification, involving the discovery of characteristics of people that we do not easily and have not yet noticed, then anything that we discovered would have to be such as to provide an independent way of making just those judgments of identity that we now make, otherwise we would have insufficient reason for saying that the heretofore unnoticed characteristics were criteria of personal identity, and not of something else, or of nothing at all. This would entail that when as in the case of Ezra, although there is no missing evidence, we do not know who a person is, the newly discovered characteristics could not show me who he was either. For if they showed him to be Peter, Paul, or even Herbert, that would prove that the new system was defective, in that it did not yield the same judgments of identity as we would otherwise make. Any new system must, to be justified, leave all the questions undecidable that are undecidable prior to its invention.

Moreover, the heretofore undiscovered characteristics would surely have to be either physical or personal. If they were physical, and some of them belonged to the brain and some to the rest of the body, then the set of charactersitics that had constituted either Peter or Paul would be disrupted by the transplant, and we would be faced with just the same quandary as faces us, given simply Paul's body and Peter's brain. If they belonged only to the brain or only to the rest of the body, they would only add a further dimension to the problem of who Ezra is: as well as asking whether Peter's memories, tastes, and affections in Paul's body adds up to Peter, Paul, or who, we would have to ask whether Peter's (or Paul's) identity characteristics, in Paul's body with Peter's personality, adds up to Peter or Paul; and this question would be no more tractable than the others.

If the new identity characteristics were personal rather than physical, that is (if this is conceivable), some new items of the same kind as knowledge, taste, wit, intelligence, the same breakdown would result; they would either be split by the transplant or would survive with the surviving brain or the surviving body. If they were split they would not show Ezra to be either Peter or Paul; while if they survived with Paul's body or with Peter's brain, we would have no better reason to be impressed by their survival than by the survival of any of the characteristics that we have all along been accepting as constituents of the problem.

It should by now be clear that the only reason we have for supposing that there is something, whether a soul or anything else, whose nature is such as decisively to settle questions of identity is the supposition that when we identify a person, we are saying that there is something about him that has not changed. Of course, when we are right *there is* something that has not changed: his identity. We say he is Joseph Mulholland and he is Joseph

Mulholland; but to say that a man's identity has not changed is to say nothing whatever about him. We do not say even of Caesar, constant as the northern star, that his identity never changes: not because identity is in fact a property of people that never does change, so that there is no occasion to mention constancy of this kind, but because if it is possible for one person to become another, it is not the same person who is at one time (for example) Caesar, and a little later Cleopatra. If Caesar becomes an alcoholic, it is Caesar who is now addicted; but if Caesar becomes Cleopatra, it is not Caesar but Cleopatra who is now Cleopatra. The impossibility of Caesar's changing his identity is not the impossibility of his becoming Cleopatra (though that too may be impossible), it is the impossibility of our calling the person he has become 'Caesar' if we accept the change of identity. In a sense identities can change, but if they do we cannot attribute the new identity to the old person.

III

At this point a perspective might be suggested that would show, so to speak, *where we are* when we ask such questions as that as to the identity of Ezra. We are inclined to think that although there is a certain complexity in the art of making judgments of identity, still *what* we are thereby ascertaining is something comparatively simple. Perhaps it is difficult to ascertain because being hidden, we require complex detection methods; or perhaps it is not hidden exactly, but any alternative to an intuitive recognition of it that we may fashion turns out to be quite complicated, in the way a set of directions for singling out an object may be complicated, although the person giving the directions may see the object from where he stands, and does not himself need the directions for singling it out. If I am trying to direct someone's attention to a building I can see from my window, I may say 'You see the group of low buildings just to the left of the tallest of the white buildings? The one I mean is in that group, and it is the one with the steeply sloping roof and the gables.' To make use of these instructions, the other person will have to decide which of the visible buildings he would call white, and which is the tallest of these; then amongst the buildings to the left of that one he will have to decide which buildings form a group, and are low; and amongst these he will have to distinguish those that have steep roofs, and of these which one has gables. That done, however, he will be able to focus his attention directly on something, and this in turn will put him in a position to devise his own ways of directing other people to that object.

We may suppose, I am suggesting, that it is this way also with judgments of identity: it can be complicated, but there is something essentially simple about it. It is simple once we have fastened on to the entity or state of affairs to which we are directed by complex instructions; but even then it can be difficult to direct another person to the same phenomenon, or to follow directions ourselves to other phenomena of the same type.

We might suggest an alternative picture of the business of making such judgments. It is indeed a complex affair; but there is no simple secret to it. We do not, somewhere in the course of being introduced to its complexities, grasp the essential point towards which it is all directed, which point, once grasped, makes the complexities seem routine and to be expected, and puts us in a position to develop for ourselves further niceties. Certainly there is a point, if not an exact one, at which we understand, at which we achieve mastery; and certainly thereafter the complexities will seem routine rather than strange and difficult, and we will be generally successful in applying what we know to new cases; but it is not necessary to explain this competence as being due to having seen something that the instruction points to but does not contain – due to having hit on the secret. This is what mastery means, for if you find the procedures strange and muddling, or if your competence is confined to the cases that have been canvassed in the instruction, you are still a learner. And it is quite to be expected that even without any sudden insight, in the course of time complicated and strange procedures will become familiar and routine.

People generally learn to use language without concerted explicit instruction; but if we were to set about to teach someone to use the words 'same' and 'different' in a few sittings, we would perhaps do it with a course of examples arranged according to a plan. Various plans might do equally well, but one such plan, which might or might not itself be explained to the student, would be that of setting out examples of the different ways in which 'the same' is contrasted with 'different': illustrating the way we say that the same thing can have different properties at different times, the way having the same properties at different times is consistent with being either the same thing or a different thing, the kinds of case in which although two things are 'the same as one another' we say they are 'different things,' and the kinds of case in which their being the same as one another makes them the same thing. If we were to explain the plan behind our course of examples we might say, for instance, that we have one way of applying the words 'same' and 'different' to individuals and another way of applying them to their properties. We might point out the different kinds of expressions we tend to use in the different cases: the way we generally say 'the same' and 'different' *simpliciter* where it is a question of difference of property, but we say 'the same x (the same man, tree, book)' when it is a question of the identity of individuals. It might then be advantageous to point out that 'the same x' is nevertheless used in cases of property differences ('same colour', 'same shape'), but that when it is used this way we can say 'almost' or 'exactly' or 'not nearly the same x,' whereas an individual is either the same x or not, and never almost the same or exactly the same. We do, however, we might go on, sometimes use expressions like 'same (or different) person' to indicate sameness or difference of property: we say that Harry is quite a different person since his operation, and by this we do not mean that he is no longer Harry, no longer the man Alice married, but that his personality has changed. He is more cheerful now, or more studious. Further, we might point out some ways in which change of property can give rise to change of expression for identity of individuals: 'same child' is generally an identity of individuals expression, but when a person has grown up he is no longer the same child, although not a different child either, but the judgment of identity now requires the use of some broader term, like 'fellow' or 'person.'

Such discourse is fairly abstract, and it would be unlikely that anyone could listen to any amount of it and then proceed to use these words acceptably. It is not just that it is too complicated to grasp and apply without practice (so that perhaps a genius could do it), but we would scarcely be able to avoid using the words 'same' and 'different' in the explanation of their use, so that some comprehension of these words would be required before the explanations would mean anything. Moreover, every general rule that we might suggest is subject to various exceptions, and there is an enormous practical problem of specifying just when the exceptions apply, while at the same time there is no particular problem, for a practiced language-user, given any particular case, of knowing what we would or would not say. We would therefore teach primarily through examples, and through getting the student to use the words himself in a variety of cases, correcting him when he goes wrong, drawing his attention to differences between one case and the next, and testing him with further cases to see if he has understood the difference we meant, and the difference it makes.

Somewhere along the way it will (or it will not) come about that the learner uses the words much as we all do, and then we will say that he understands them, or that he knows their meaning. At that point he himself will be likely to have the familiar feeling of knowing the meaning of the words, a feeling that they are like old friends to him, a confidence that he can offer any number of clear cases of things being the same or different, and perhaps a feeling that he knows something that enables him to see that these are indeed clear cases, and that enables him to explain the words to someone else, not necessarily just by repeating the explanations by which he himself has learned them. But clearly it is not necessary to suppose that he knows something over and above what he has been taught. All that has happened is that something that was initially strange and difficult for him has become familiar and routine. To account for his ability to explain the matter in new ways, if he has that ability (and not everybody will), we need only suppose a general talent for explaining things: a talent, for instance, for analytically surveying what he does, for diagnosing where the other person's difficulty lies, and for devising ingenious ways of surmounting such problems as that posed by the fact that a learner will often not understand the words that one is initially inclined to use for explanatory purposes.

Now if, instead of supposing that when a person's performance in using the words 'same' and 'different' has reached an acceptable standard he has hit on something not contained in the instruction, which enables him to proceed independently, we suppose simply that he has mastered the complexities, what we are therein supposing is that he now proceeds as we do. This will entail, among other things, that he has generally no doubt in the cases in which we have no doubt, and is generally uncertain in the cases in which we are uncertain. Mastery, that is to say, is not a state in which one can answer all possible questions, but one in which the answerable questions are answered correctly, and the unanswerable ones are not answered. If this is so, we will have no reason to expect that in the problem cases there must, if only we could hit on it, be an answer to such questions as whether this is the same x or a different one.

Suppose, to illustrate, that late on in the course of instruction such as has been described we take up with the student the case of Peter, Paul, and

Ezra. We would watch him with great interest to see what he would do with it. What might make it difficult for him is that he may have become accustomed to expect that whatever he does, we will be able to tell him if he is right or wrong. He may not be prepared for our saying that he would be wrong whichever way he answered, not because of some subtle feature of the problem that makes us say that Ezra is someone other than Peter or Paul, but because we are teaching him what we say in various cases, and there is nothing established as to what to say in Ezra's case.

Something like the position of this language-learner is what we are all in when we ask some sorts of philosophical questions. We have come a long way in mastering certain words, and then a question comes up that is interesting because answering it seems to involve going beyond anything we have so far learned. We as it were take such a question to be part of an advanced programme of instruction, a question to which there will be, as usual, an answer. What we overlook is that we ourselves by now, if anyone is, are members of the community of masters, and if we do not know what to say, there is no one to serve as our teacher and show us the way. There just is no answer.

Should we also say that there is no question? An *apparent* question arises naturally enough, as we saw, from the way we routinely settle some aspects of some questions about marital status, criminal responsibility, contractual obligations, and even (as in the case of a mother's long-lost son) ties of love and concern, by establishing the identity of a given individual. If the person before us is Peter, then he is legally married to Mary, he is liable for any offenses with which Peter can be charged, he has obligations under this or that contract. In many circumstances, we only need to know who he is. It is natural, therefore, to expect that establishing the identity of Ezra would immediately untangle problems created by the brain transplant.

Can we say, however, that though it is natural to ask who Ezra is, it is still a mistake? If we are right, it is a mistake to give a yes-or-no *answer* to the question whether Ezra is Peter; but is it, therefore, a mistake to *ask* the question? The question whether to ask the question leaves one uncertain in the same sort of way the question itself does; so perhaps the answer is that it, too, is something about which nothing is settled, and hence is not answerable.

Imagining

Philosophical offerings on the subject of imagining have not so far included
any very clear alternatives to mentalism on the one hand or behaviourism
on the other, although there is ever-increasing documentation of the
unsoundness of both kinds of theory. In this essay, while I may make some
additions to that documentation, my main ambition will be to offer a
tenable account of imagining which is neither behaviouristic nor
mentalistic.

We use the word 'imagine' in a number of different ways, and as a
prelude to indicating the kinds of imagining to which my claims apply, I
will describe some of the uses of the word that I will not be discussing.

In the first place, 'imagine' is used sometimes in the expression of
enlightened guesses, for example when we say, 'I imagine she will be quite
pleased.' There is little inclination to suppose that a person who so expresses
himself would, in another sense of 'imagine,' be imagining her pleasure.

Secondly, we use 'imagine' sometimes to express our opinion of the
baselessness of what someone is thinking or saying. 'Oh, you are just
imagining that!' is a way of saying something like 'That is not true, and
you would not say it if you were being careful to confine yourself to what
is true.' Here, it may be suggested, we are capitalizing on what might be
called the truth-exemption feature of another use of 'imagine': in a large
class of cases, to imagine is partly to ignore considerations of truth or even
of probability; and hence in saying 'You are just imagining it' we make

as if, when what a person says is not true, he knows it is not, the way one does in these other cases. Here the 'imagining' designation arises entirely from the hearer's belief that what the speaker says is not true: without the speaker's attitude or purpose being thought to be different, he would not be described as imagining if what he said were taken to be true.

A case, thirdly, which although quite similar in some ways to the last-mentioned, is more easily confused with those I wish to discuss, is that in which, for example, I am late getting home and my wife says she has been imagining terrible things happening to me. Here, I suggest, she uses the word 'imagine' in her relief at the sudden recognition of the baselessness of her fears. As long as she did not either believe or fervently hope that the thoughts she kept having were groundless, she would not describe them as imaginings. If she were talking to a friend while still awaiting my return, she might say 'I just know something dreadful has happened. He has been in an accident, or has had a heart attack'; and only in moments of calm when she is no longer much inclined to believe such things will she say that she keeps imagining terrible things. The use of the word 'imagine' here is a way of saying that a thought is not to be taken seriously.

There is a fourth, and somewhat problematical, kind of case that we will not be discussing, the case in which in idle moments or on summer days the mind runs on, more or less aimlessly developing trains of thoughts or pictures, whether pleasant or painful. I lie in a hammock and soon I see myself in my mind's eye sailing through the inky blue Caribbean. I can almost feel the motion of the boat, and its eager response to gusts of wind. Is this a case of imagining? I think many people would regard it almost as a paradigm; and yet I am not certain. If I ask myself in what kind of circumstance, outside of pholosophy, I would so describe it, it seems clear, in the first place, that while I might later report that I spent some of the afternoon day-dreaming, I would not say that I spent some of the afternoon inagining. I might, if the case were only somewhat different, say that for a time I imagined I was sailing the Caribbean; but probably only if the experience had been so real that I was surprised or disappointed on suddenly finding that all the while I was on the hammock in the garden. The case would then resemble our third kind of case.

The cases, now, that I do wish to discuss are either (a) cases in which we set out to do something specifically as imagining, for example, cases in which we are asked to imagine something, and we do; or (b) cases where what we do, under whatever auspices, is so recognizably similar to what we do specifically as imagining that we can be confident that in so describing it we are saying something about *it*, rather than using the word 'imagine' only because of the way things have later developed, and to express our relief, or our ambivalence as to whether to take seriously the thoughts or visions we have had. The main examples I am able to suggest in this second category are the creating of such 'works of imagination' as stories and plays, some of the activities generated by reading or hearing works of imagination, and some of the activities of children at play.

I want, in short, to talk about cases of imagining that remain so describable however things turn out, and whatever our attitude or emotional state is. Because they are the clearest cases, I will mostly use as my model what we do specifically as imagining: cases in which we do something in response to a suggestion that we imagine, or cases in which we say 'This is how I imagine it.' In doing so I do not wish to imply anything more than that these are the clearest cases.

What is imagining? The idea that first presents itself, and which will haunt us throughout this essay, is that imagining is generating a mental picture, or a train of them: that is the one thing we must do by way of imagining. That being done, we have imagined. We may then describe what we have pictured; or children may, as they picture something, act out the part of the princess or the ogre they are picturing; but describing or play-acting or drawing cartoons or anything of that kind is, according to this account, ancillary to or parasitic upon the imagining itself – the picturing. In expressing this view, some people will and some will not say that they mean to include amongst 'picturable' things for this purpose, sounds, odours, flavours, feelings which cannot be captured in ordinary pictures.

If we were to agree that imagining is, in this broad sense, 'picturing,' some of the philosophical questions that might arise about imagining would be as to the exact character of this picturing. Are the pictures like a picture on a screen except that they have no edges and the screen does not remain when the picture is removed? Do they seem close to us, or far away? Can one have a closer look at them and thereby discover unsuspected facts about them? Can one imagine with one's eyes open, and if so do the pictures appear in amongst the things that one sees, or if not where do they appear to be? If I can draw a good imaginative picture of a fairy castle, what role does picturing play here? Would I in drawing it be copying from a picture, or would I, in either drawing or picturing it, be employing some pre-pictorial ability? And may my picturing at times seem satisfactory to me, not because I achieve a good likeness, but just because I have no doubt that I know very well the appearance of the thing pictured?

Some ways of answering some of these questions might be pressed by someone who was sceptical about the very existence of such mental processes as picturing. It might be suggested, for example, that people, when asked to imagine the Parliament Buildings burning, may fancy that they picture it faintly, just because they do not doubt that imagining is picturing, and they are in no uncertainty as to the appearance of the Parliament Buildings. When pressed, the believer in the faintness of mental pictures may say that that is just the way imagining is. It has nothing like the rich and strong quality that perceiving has; and it is quite to be expected that the mental picture of a brilliant yellow cow should not itself be brilliant yellow. The sceptic may then press the idea that it is very odd in that case that so many of our accounts of how we imagine things should be drawn in terms of definite outlines and strong colours.

I have no wish, however, to deny that picturing occurs, and it seems to me an interesting and possible project to describe its peculiar phenomenology. My interest in picturing is confined to whether, if it exists, it plays either the kind of role in imagining that I described, or any other essential role.

Whatever may be the precise nature of picturing, then, do we imagine by picturing? Suppose that when someone is asked to imagine the arrival of his mother-in-law for a visit, he says he would imagine it this way: 'She would come into the house all smiles and laden with little things for the children, and would greet me with excessive cheerfulness as if to say that past tensions were forgotten ...'; but when asked (with whatever explanations might be necessary as to what was meant) whether he had pictured the scene, he replied 'No, it just right away seemed to me that was how it might go.' We would surely not in such a case say 'Then you haven't imagined it.' Describing a scene is a perfectly regular way of imagining. At the very least, therefore, we will have to say that we imagine either by picturing or by describing. Describing is a way of imagining; but we do not yet know whether picturing is.

There are some things, for example, odours and pains, (i) that cannot, in a sense which will be explained, be described, and (ii) that some people at least cannot picture. Can anyone who cannot picture these things not imagine states of affairs in which they figure?

Before answering this question I should enlarge on the claim that we cannot describe such things as odours – for example, cannot describe the building smelling of coffee. There are at least three descriptive performances that we could give in such a context; but I will argue that none of them quite serves as imagining the building smelling of coffee:

I 'The building smells of coffee' is itself a descriptive utterance; but in saying this we would be doing no more than accepting the suggestion that we imagine the building smelling of coffee. We could dispense with this description. It is already supplied in the suggestion. To accept a suggestion of this kind is, as we will see, a way of imagining; but it is not a way of imagining by describing.

II The use of such descriptive adjectives as 'heady' or 'haunting' might appear to take us beyond simple acceptance of the suggestion; but in using these adjectives we would not be doing quite what we were asked to do, any more than if we described the building as smelling of Columbian coffee we would be just exactly complying with the request that we imagine the building smelling of coffee. If this form of descriptive supplement were permissible, then a new question would simply arise: how is it possible to imagine the building smelling of *Columbian* coffee?

III If we described the way some people went about breathing deeply and looking delighted, while others were annoyed and complained of not being able to get on with their work, we would not be describing the building smelling of coffee, but describing a consequence of that supposed fact.

Given then that I can neither picture nor further describe some states of affairs, can I nevertheless imagine them, and if so, how is that done?

My suggestion is that we *can* imagine in such cases, and that it is done by agreeing to the proposal that one imagine something, and going on from there. Someone says 'Imagine the building smelling of coffee,' and I do not see what he is getting at, but I say 'All right, what then?'; or I think I see and I say 'Yes, some people would go around looking delighted and breathing deeply ...' In the one case he goes on from what we have imagined, and in the other case I do; but in neither case is what we go on to the imagining itself. In both cases imagining the building smelling of coffee is accomplished just by the stage-setting agreement that we reach.

Someone might want to say that although we may agree to imagine it, we still have not done so. We agree to do it and then we do it; and the question still remains, *what* do we do?

If I am asked to imagine that these children who are laughing gaily are in pain, I cannot do it by supposing them to be crying and inconsolable: so *what* do I do? The answer to that, as Wittgenstein suggested (PI §393), is that I imagine them to be in pain. I might go on from there to say that they must be extremely good actors, to feign gaiety so successfully when they are in pain, and that surely they couldn't keep it up for very long; but again saying these things would not be part of what it is to imagine them in pain, but consequences of so imagining. What is that imagining itself?

One may be inclined to suppose that we have to picture to ourselves the sensation the children are having; and perhaps also that it will be from this picture that the imaginer will understand how brave the children must be, what good actors they are, and so on. Yet if we did have a picture of pain when we imagined it, it would not be from the picture that we would see how brave and artful the children must be; but we would *read into* the picture what we know about pain. It would be from our understanding of the concept of pain, not from the picture of pain, that our further conjectures about the children would be drawn. And if that is so, we do not need the picture. If we understand the concept we can imagine that the children are in pain just by saying 'All right, what then?'

We are not inclined to deny that one can *suppose* something just by going along with the request that we suppose, for that is the prime way we have of supposing; and yet what is the difference between imagining and supposing?

'Imagining that' and 'supposing that' share many characteristics, and in many contexts may be used interchangeably. They are akin in at least the following ways: (i) to 'do' either one we do not need to add anything to the suggestion or proposal that we do so; (ii) to suggest that we do either one is among other things to suggest that we waive considerations of truth or probability; there is nothing *logically* offensive about supposing or imagining what is false or improbable (even if we may sometimes wisely decline to do so), while there *is* something logically offensive about supposing or imagining what is true. I couldn't, in 1972 (given rudimentary political awareness), either suppose or imagine that Pierre Trudeau is prime minister of Canada. (iii) Both supposing and imagining are the beginning of something: having either supposed or imagined, we go on to develop consequences, but the development of the consequences is not itself the supposing or the imagining.

They differ in at least the following ways: (i) we would generally prefer 'suppose' in more formal, more serious proceedings, and 'imagine' in more casual and conversational contexts; (ii) 'imagine' is preferable for extremely fanciful suppositions: one baulks a little at supposing that a cow jumps over the moon, but hardly at all at imagining it; (iii) one supposes but does not imagine that p $\supset$ q, whereas one may either suppose or imagine that one's mother-in-law arrives for a visit. The use of 'imagine' seems to be confined to cases where the thing imagined is picturable or describable, even though the imagining of it does not require that it *be* pictured or described. And (iv) the development of consequences in the case of supposing is in a broad sense logical: one makes inferences from the supposition and other things one knows; whereas the tendency in the case of imagining is to develop the consequences (one might say) empathetically. Whereas the consequences of suppositions are reckoned and argued over, the consequences of imaginings are intuitive. If I fail to see a point of yours in an imagining context, I can be made to see it not by argument but by artful description, designed to sharpen or enrich my intuitions.

I have stated these contrasts too baldly; I simply want to suggest that the use of these words gravitates in these directions: that being more akin to one model or the other is a reason for preferring one word or the other. But there are few contexts in which it would be in order to use one of these words, and quite wrong to use the other.

Because of its close resemblance to supposing, and because doing it involves adding nothing to the proposal that we do it, one might be inclined to treat *imagining that* as a secondary or degenerate or in some other way uninteresting case of imagining, and to say that the mainline cases are those in which 'imagine' takes a direct object. Personally I find *any* use of a word interesting, and therefore I am not inclined to say this in general, and still less in the case of *imagining that*, since (i) this construction is very frequently used, and (ii) (as I think I have shown) it has a life of its own and is not, for example, just a careless way of saying 'suppose that.'

Imagining, though not the same as supposing, is not different in any way that would require us to say that although one of them does not, the other does involve doing more than going along with the suggestion that we imagine or suppose.

Both imagining and supposing require that we should understand the proposal, but that is not something we do (usually very quickly?) before anything else can be done. It is true of us *before* we are asked to imagine or suppose something that we do or do not understand the suggestion.

We do not first hear a suggestion that we imagine something, then understand it; but rather we hear and understand it, but neither simultaneously nor successively: one might (somewhat desperately) say 'we hear it understandingly.' Similarly, at least in the case of *imagining that*, we do not agree to imagine, then imagine, then reflect upon what we have imagined. It does not take time, even a very short time, to get the imagining started or finished; people do not compare notes as to how quickly they can imagine such things as that the children are in pain, and are not amazed at the speed with which some people can do it; and we do not enquire, as the conversation continues about what we have imagined, whether the other person is still imagining, or just managing on the recollection of having done so. Nor do we check into just what a person has done by way of imagining the children to be in pain, to see if he has done an adequate job of it.

It may seem surprising that imagining could have so little *volume*: that, for example, imagining different things should be done in the same way, and that no effort of invention should be required. But I think that what I have said about the role of understanding explains this. Imaginings of this kind are not self-sufficient, but are the beginning of something; we go on from there to amuse ourselves with or otherwise explore the supposition we have made. But to make the supposition we need only understand the imagining assignment.

In discussing 'imagining that,' I have stressed cases involving smells and pains, where most of us do not claim to be able to produce mental likenesses; but I do not wish to imply that it is only in such cases that 'imagining that' has no 'volume' and is 'done' by going along with the suggestion that we imagine. I have stressed this kind of case only because the difficulty here of producing a mental likeness helps us to see that we can indeed imagine something without doing anything whatever. But this is just as true of any case of 'imagining that.' If you ask me to imagine that Charlie Chaplin is prime minister of England (as distinct from imagining him as prime minister), I may but I need not picture him entering 10 Downing Street, or slouched on a front bench in the House of Commons, or anything else: I can just say 'All right; what then?' and you would not exclaim 'What! Have you done it already?' or insist on knowing *what* I had done so quickly.

I have no wish to say, however, that all imagining is as slight an affair as the kind I have been discussing. For if, when asked to imagine myself having lunch with the Queen, I describe a scene full of conversational false starts, awkward silences, and misunderstood witticisms, these and such things would *be* what I imagine, and could not be set aside as the consequences of imagining it. Whereas in the other kind of case one does nothing by way of imagining something, in the cases in which 'imagine' takes a direct object, one would do a good deal.

It seems, then, at least roughly true that when asked to imagine *that* something or other (for example, imagine that your mother-in-law were to come for a visit), nothing is required but that we should understand and go along with the proposal, whereas when asked to imagine such-and-such (for example, the arrival of your mother-in-law), some contribution of content is necessary. We might call these kinds of case 'indirect object' and 'direct object' imagining respectively;[1] and a further division of the latter kind into 'short prescription' and 'extended prescription' imagining will be useful. When what is to be imagined is designated just by a proper name or a definite or indefinite description, for example, 'Imagine your grandfather' or 'Imagine the Peace Tower,' it will be called 'short prescription direct object imagining,' while when we are asked to imagine something or other about someone or something, for example, 'your grandfather dancing the Watusi' or 'Mackenzie King's reaction to the miniskirt,' it will be called 'extended prescription direct object imagining.' I will have a few things to say about the former kind of imagining at the end of the essay, but most of the remainder of the paper will be given over to the latter, it being by far the more common and philosophically interesting. For convenience, I will refer to the direct object extended prescription

1 The essential distinction here is between imagining that can be 'done' just by accepting the suggestion, and imagining that requires some invention. It is very generally, but perhaps only very generally, true that in cases of the former kind 'imagine' takes an indirect object, while in the latter cases it takes a direct object. An obvious exception is 'imagine the building smelling of coffee' which, although a direct object construction, seems clearly to be accomplished just by accepting the suggestion. It is not clear to me whether we need a sure guide to the different cases; but if we do I think it will not be grammatical, but will have to do with what we might call the social situation in the different cases, or (one might say) with what kind of game is being played. When a person has a point he wants to make, and asks one to imagine something as part of his procedure for making that point, all that will immediately be required is acceptance of his suggestion; while when he needs help in imagining or understanding something, or when he wants to explore the other person's wit or perceptiveness, invention will be required. Yet, if that is along the right lines, it is still not enough, if only because not all cases of imagining are cases of being asked to imagine something.

imagining simply as 'imagining'; and I will make seven main points about it.

1 It is not an activity. One might think that since we can be asked to imagine something, and can comply or fail to comply with such requests, imagining is surely something we *do*. But if so, there would surely be a time when the activity of imagining is under way. When would that be? I can suggest only two possibilities: (1) that we are imagining when or if we reflect on or prepare what we will do or say by way of complying with a request to imagine, or (2) when we deliver, whether publicly or privately, what is to serve as compliance with such requests. Both the delivering and the reflecting are activities, but neither, I will argue, is what we would call a case of imagining something.

1 If someone asks us to imagine something, we are certainly likely to be quite busy for a time, considering how it might be, or what we should say; after which we may say 'I imagine it this way' and then set forth our account. We perhaps entertain various pictures and developments, are pleased with some and dissatisfied with others, and finally decide how it might go. What we sometimes do prior to saying how we imagine something is certainly an activity; but it would better be called 'trying to imagine' or 'wondering how we would imagine' than 'imagining.' If this were the imagining, then saying what we imagine would appear to call for a faithful account of what we have been through in (as I call it) trying to imagine. But this of course is not what we do. Not only do we omit the unsuccessful ideas we have had and rearrange and modify the successful ones, but it may suddenly strike us how we can do better than even the best ideas we have had. And this 'being struck' need not consist in suddenly having a new idea appear: we may just at some point find ourselves able to proceed confidently on a line of development which is in fact unlike anything we have been contemplating. But if what we finally say as to how we imagine something is different from anything we have thought of in the course of trying to imagine it, it would not be said that we were *lying* in saying the new idea is how we imagine it.

The fact that we say 'This is how I imagine it,' rather than 'This is how I have just imagined it' or 'This is how I am imagining it,' is of some significance here. We do not speak as if we were reporting on what has just now happened, or on what is currently happening, but rather we regard ourselves as *pronouncing upon* how, for our part, it is to be imagined. If someone misunderstands our verbal or other sketch of an imagined scene, we will more naturally set him right, not by reiterating the sketch, but rather by adding freely to it, or by going about explaining in quite a different way. These further explanations will be derived neither from a public sketch nor from a private one, but just from the imaginer. It is as if, from this flesh and bones, something charming appeared by magic.

It would be possible specifically to request that a person privately entertain various pictures and possibilities as to how something might be imagined and then report exactly what he has done, but that would be very different from asking him to imagine something. It would express an interest in the phenomenology or psychology of imagining, and not in, for example, his vision of the scene when his mother-in-law arrives – that is, in the story he could make of those ingredients.

2 One might be driven by such considerations as these to say that imagining is what we do when we finally say or otherwise show how we imagine something. But this I think would be wrong too: we do not prefix these performances with, for instance, 'I am about to imagine,' or 'I hereby imagine'; and we do not say we have finished imagining when we finish saying how we imagine something.

We say how we imagine something, and while this may look a little like a report on what we have done privately or on what we sometimes or frequently or habitually do, there is nothing that counts as a case of our actually doing it.

11 It is not the possession of specific skills or powers that enables people to imagine things. Imagining is not a practice indifferently teachable to anyone with the required intelligence, but a form of personal expression.

If I am unable to throw a ball fifty yards, it is perhaps because my muscles are too weak, or my stance is poor, or I keep my arm too straight; and by examining my style it could often be predicted what my ball-throwing capabilities would be. But there seem to be no corresponding properties of a person enabling us to imagine things.

There are, it is true, habits and dispositions that over the long run will make a difference to how people imagine things. If I have a habit of noticing and remembering curious features of the passing scene, or of juggling the constituents of situations to produce interesting variations, or if I have the kind of detachment that prevents me from being so involved in an incident with my mother-in-law that I fail to remark nuances in her behaviour and in my own, I will be more likely than otherwise to have something interesting to offer when asked to imagine her arriving for a visit. Such attributes, however, are not techniques of imagining. I do not imagine her arriving *by* being detached or noticing curious features of the passing scene, the way I take a good golf shot by standing in a certain way and keeping my eye on the ball, but if I have those habits, when asked to imagine something rather more interesting things may occur to me than otherwise. Such things, however, will simply occur to me, or not. There is nothing I do to bring about their occurrence, and *a fortiori*, nothing that, if done with *finesse*, will bring about the occurrence of more interesting things.

There are two kinds of criticisms of a golf performance: remarks about what the ball does, and remarks about what I have done that brought this about. We say 'Too high; stand a little further forward'; but about a piece of imagining we make only what corresponds to the former criticism: we say it was not very perceptive, that it lacked richness of detail, and do not venture opinions as to why it is that way.

In general it is possible to develop techniques for doing something quite definite, for example, for making a golf ball veer in flight to the left or to the right; but imagining is not something definite in that way. If I find myself unable to say how I would imagine a meeting between myself and the prime minister, my hesitation will not necessarily or even likely be due to my inability to describe his appearance, or some place and occasion when we might meet, or some possible lines of conversation between us. It is not delivering something that is difficult, but delivering something *imaginative*. We hold our tongues and prefer to say nothing when we can find nothing to say that is charming, or perceptive, or that makes a good story. There are, however, any number of possible inventions that would be charming or perceptive, and hence imagining is not a sufficiently definite enterprise to permit our working out, as we may for an intentional slice in golf, what actions of ours will achieve it. It is not yet ascertained, when one is trying to imagine something, what success in the enterprise will be like.

Although, as was suggested earlier, we may in some ways study to be more witty, perceptive, ingenious, when the occasion arises to be witty, one does not employ some ability perfected over a period of time, but with or without a little reflection one simply says something, and it is imaginative or not, as the case may be. Cultivating one's ability to imagine is not like practicing one's golf swing, where when the day comes to play golf one does just what one has been practicing. Rather it is as if the way to be a better golfer was to practice yoga exercises, and people found that afterwards without half trying they played superbly.

III To say 'It was just as I had imagined' or 'just as I had always imagined' is not to refer to acts or events or processes of imagining that have occurred in the past.

If I see the film of a book I have read and say 'It was amazing! It was all just as I had imagined it when I read the book,' does this mean that as I read the book I conjured up pictures of the scenes described and of the characters in the story, and that I remember these pictures, and there is a remarkable resemblance between many or all of them and the scenes and persons in the film?

Well, perhaps I did do such picturing, and perhaps also I remember it all clearly now, although it is a little surprising that we should remember these images so clearly when we have difficulty often remembering persons and things we have actually seen. And there are puzzles too as to what are the mechanics of the comparisons we are supposed to have made: are the recollected image and the film we now see both present to us at the same time, and if so do they overlap, is one superimposed on the other, or are they seen side by side, or do we see only one at a time and switch back and forth from one to the other? And if there is a present image with which we compare part of the film, how do we know that it is like the image we had when we read the book? A further image with which we compared *it* would serve no purpose, because the same question would arise about *this* image.

It might on the other hand work this way: I understood the book in a certain way, and given this understanding of it I would have chosen just these actors for these parts, would have designed the sets in just these ways.

What is it to understand a novel or a play in a certain way? Well, suppose it is in part, for example, whether you take Polonius to be a wise old patriarch or a platitude-muttering old fool. You will then be satisfied or dissatisfied with the way he is played on stage or in a film. But 'taking' him in a certain way is not necessarily or even generally having a certain mental picture of him: it *may* be a matter of muttering 'Old fool!' and such things to oneself as one reads the parts of the play in which he appears; or of thinking 'Isn't he like old so and so?' or of feeling a certain aversion or a certain affection for him. But it need not be any of these things: it may just be that one is dumbfounded later on when one hears him described as a wise old fellow, or hears his advice cited approvingly. This is to say, it may be that there is absolutely no puzzle for me as I read *Hamlet* as to what sort of a man Polonius is. It may be so utterly straightforward that the script itself tells me, with no interpretation, that he is an old busybody; and if I react to him as I read, it is not a reaction to an interpretation but to the character just as he appears in these pages.

Now, if my taking a novel in a certain way were to consist, in the above manner, phenomenologically entirely in the fact that at various later junctures I agree entirely or disagree strongly or have my doubts about various representations that are made about it, this is just the kind of circumstance in which I would be likely to say 'That's not the way I imagined it,' or 'That's exactly the way I imagined it.' Having understood the novel in a certain way, the film *now* strikes me as all wrong, or just right, as the case may be.

What I am suggesting might be explained somewhat differently this way: when I see a film of a book I have read, it is as if there is a project of joint imagining in progress, like but ever so much more extensive than the little projects we have mostly been discussing, but in which the assignment was given some time ago, namely, when I read the novel, and the film plays the same role as another person making suggestions as to how it might be imagined, which I like or do not care for. At each stage it is as if the film puts to me the question, 'Would you imagine it this way?' and depending on how I have understood the assignment, namely, the novel, I answer 'Yes,' 'No,' or 'Perhaps.' If the film is 'Just as I imagined,' the answer all along will be 'Yes.' But this will not show that when I read the book I experienced some images which I now remember, but only that from my understanding of the book I accept the film as an altogether suitable representation. It will be *like* the case in which someone says 'Let's imagine such-and-such,' and we each contribute parts to the story, each accepting the parts contributed by the other, except that in this case the film does all the contributing and I do all the accepting. But in accepting I am not saying 'You guessed just what was in my mind,' but rather 'That fills the bill splendidly.'

IV Imagining is at least very often a joint enterprise between two or more people. A great many of the examples we have been using presuppose this, yet it will be surprising if we are inclined to think of imagining as an essentially private affair. Clearly, though, the way it often goes is that someone suggests imagining this or that, and if I agree, that is the start of something, and of something that can happen only if he stays to participate. It is very odd to suggest imagining something, and then depart, and equally so, to agree to the suggestion and then depart.

One might have thought that having been asked to imagine something, one must do all the work oneself; but *this* is a case of imagining: I say 'Perhaps she would say ...' and you chuckle and reply 'And of course you could only answer ...'; I had perhaps not thought of that but I like it and I continue 'And can't you just see how she would blush and struggle to maintain her composure?' This is of course another piece of evidence that a proposal to imagine is not a request that one do something in the head. Someone at this point might say that the real imagining goes on in the head, and that what we say to one another merely helps to design and shift the mental scenery. But if this were true then it would be possible for *this* to be a case of imagining: you say 'Let's imagine such-and-such,' and then we both sit and entertain various thoughts and images, and after a time we say 'That was fun; let's do it again.'

The absurdity of this is not due only to the fact that neither of us knows what went on in the other's mind, or even whether anything went on. For one thing if I knew you well and had watched your face as you entertained those thoughts and images, I might have a very good idea what went on. But whether I know or not is not the point: the point is that in this case there is not yet the distinction between imagining and trying to imagine. Neither of us would know whether to say that we had carried out the assignment, or only that we had considered how it might be imagined. If you asked me how I imagined it, I might answer you, certainly, but my answer would not, as we have seen, necessarily or even likely be a faithful account of the thoughts and images I had entertained.

But in the case in which we jointly and explicitly imagine something, by contrast, a reconstruction later of how we imagined it would have to go in just the way our conversation developed. (This is perhaps only roughly true. If we had explicitly made amendments in the course of our conversation, only the version with which we seemed best satisfied need be given as how we imagined it; and of course since our conversation would not generally have been intended to settle anything very definitely, doubts might remain afterwards as to what version we had preferred, or as to the sequence we were supposing the events to occur in. But anyone listening to a tape recording of our conversation would be likely to have the same doubts and the same certainties as to how we had imagined it, while a mental videotape of the thoughts and images I had privately entertained would give rise to no certainties and no doubts as to how I imagined it.)

v The above considerations show, I think, that there is a delivery-of-the-goods feature about the word 'imagine.' It is the things we say, the cartoons we draw, the play-acting we do, that show how we imagine something. The distinction between imagining and trying to imagine, considering how we might imagine, and so on becomes possible only in terms of the goods we have delivered.

I do not mean that the goods delivered *are* the way we imagine something: we have just seen that we may revise our account, and may not be careful to show which of two or more versions we are satisfied with. But (a) what we deliver is *normally* how we imagine something, and (b) in the unclear case, what we have said or done provides the only basis on which such certainty or meaningful doubt as there may be is possible.

The doubt mentioned here is not a doubt as to what has gone on in the minds of the imaginers. It is not because imagining is in fact *something else*, that saying something (or play acting or cartoon drawing) is not imagining. The doubt is as to whether what they say or otherwise show is accepted by the imaginers as *good enough*, up to standard. When I am wondering whether I would imagine something in such-and-such a way, I am not trying to remember just how I have imagined it, or considering whether those words exactly describe how I have imagined it, but rather I am engaged in an effort of *invention*: I am trying to think of something better, more interesting, more perceptive.

The goods delivered are not how we imagine something; and nor is *delivering* the goods imagining. There is not, as we saw, any such activity as imagining. But the goods delivered are all that can show how we imagine something. Our willingness to deliver them shows us as much as anything else whether they are up to our standard.

We talk of 'showing how we imagine something,' and this suggests that what we say or do is evidence of something, but that that of which it is evidence is not revealed, lies hidden. This is also suggested by such facts as that we are sometimes ready with any number of alternatives as to how something might be imagined. It looks as if there were a well, from which any given thing we might say or draw is only a sample, or a warehouse into which these things only provide a glimpse. And there is something right about that; but what is behind the scenes is not not-yet-delivered descriptions of scenes or conversations, but what we might roughly call our idea of what makes a good story, our idea of what is amusing, our taste in what is splendid or pathetic. What is behind the scenes is not a warehouse but a factory. We do not requisition the goods, we custom make them, and deliver them if they are up to standard (and sometimes if they are not). We do not judge a factory by its rejects or by its not-yet-finished goods, but by what it is prepared to deliver.

VI The above illustrates in one way my sixth point about direct-object imagining, what I will call its self-expressive feature. The kind of goods I deliver show something about the kind of person I am; and this, not by showing the kind or extent of the goods I have salted away, but rather something about my taste, humour, ingenuity, and perceptiveness.

Of course anything we say or do may have this function, but we seem to use the word 'imagine' in such a way as to make a special point of it. We decline to offer anything if we can think of nothing that is interesting or perceptive; and if we think of something but it is dull, we do not regard ourselves as so imagining, but concealing it from other people, but as not yet having anything to offer that we would call imagining. 'This is how I imagine it' is not a report of my recent or current state of mind, but is what we say when we have something to offer that seems to us good enough. We are in effect saying 'Let this stand as an example of my wit, taste, perceptiveness.'

VII When we ask someone to imagine something, our use of the word 'imagine,' rather than, for example, 'depict,' 'describe,' or 'suggest,' has three main functions: (1) to indicate that considerations of truth or probability will be irrelevant; (2) to show that what is wanted is the kind of rich, co-ordinated, inter-related representation in which pictures or images excel; and (3) to invite the exercise of wit, perceptiveness, ingenuity. Let me comment further on each of these points.

1 When we are asked to imagine something, then if it is something that has happened, is happening, or will happen, we are not of course *forbidden* to tell it as it was, is or will be. But an imaginative representation is not a failure, and does not even deserve less than the highest mark, if it

turns out to be entirely untrue. When an imaginative enterprise concerns actual persons or states of affairs, one way of earning high marks is by artfully representing events as taking just those turns they would not take, or people as saying and doing just those things they would not do. Such a performance would of course rely on shrewd observation and sound understanding of the persons and events figuring in the imaginative enterprise; and in this and similar ways *perceptiveness* makes for success in imagining; but one can produce a very perceptive, and therefore an entirely successful, imaginative account of how a conversation might have gone, without getting a word of it *right*.

2 I am scarcely imagining myself married to Henry viii if I think of a duly executed marriage certificate and of myself beside him on a somewhat less grand chair in a throne room, nor am I quite imagining time reversed if I think of clocks running backwards and Kennedy succeeding Johnson as president of the United States. In proposing that anyone imagine such things we are asking for more than that, and we would begin to be satisfied only when some fairly full-blooded picture began to emerge: when breakfast table conversation, domestic squabbles, and moments of joy were described, in the former case; and in the latter case when it was spelled out whether we would grieve over the reverse-birth of a human being (since that would be the end of him) and rejoice over the reverse-death, or if not, what grief and rejoicing would mean in that world – and whether it would be after or before three o'clock that clocks read 2:30, and if the latter, what 'before' would mean in that world, and so on.

One will say 'I can't imagine time running backward' if one gets bogged down when one tries to flesh out the picture of that world in any detail, but hardly when one cannot whether verbally or mentally do something corresponding to running a moving picture backwards (although no doubt I would have some difficulty when I attempted to run its sound track backwards).

3 It was suggested that we use the word 'imagine' in part to invite the exercise of wit, perceptiveness, ingenuity; and along the same lines in various places I have suggested that an imaginative representation is one that is beguiling, or that captures a certain pathos, or that makes a good story. There are at least two problems to which these claims may give rise:

i Can it be *part of the concept* of imagining that it should be ingenious, perceptive, entertaining, or should imagining not be defined without using such terms? (In the latter event we could say that people appreciate imagining when it has some of these qualities, but that something may be imagining whether it has them or not.)

ii If indeed notions like perceptiveness, wit, and ingenuity are part of the concept of imagining, must they not be included in some formal way? For example, must it not be possible to specify somehow the conditions under which wit is appropriate and inappropriate in imagining; or again must it not be possible to specify how witty, how ingenious one must be in order to be imagining?

Our discussion of these two questions will I think bring out ways in which the concept of imagining does not fit some standard philosophical models for conceptual analysis; but this may only show the inadequacy of these models.

i There are such things as cars, and people like them if they are fast, comfortable, quiet, but something may still be a car if it is noisy, slow, and uncomfortable. Again, swimming is propelling oneself through water, and people admire it if it is fast or graceful, but it is still swimming if it is slow and awkward. Should imagining not be analyzed along the same lines, and hence defined independently of its wit or perceptiveness?

It is extremely unlikely that there should be only one word that has value concepts built into its definition, but *a priori* quite believable that some words should and some should not. It would therefore be desirable if we could show that 'imagine' is not alone in this regard.

I think there are in fact quite a number of words or expressions that resemble 'imagine' more than they do 'swim' in this way. Would one say that playing the violin was the production of noises with that instrument, and that people liked them if they were rhythmic or in tune, or would one say that making noises with the violin was not 'playing it' until it achieved a certain musical standard? (We would perhaps not deny that Jack Benny plays the violin, but how much worse than that could a person be and still be regarded as playing the violin?)

'Imagine,' it seems to me, is more like 'play the violin' or 'dance' than it is like 'walk' or 'swim' in this regard. The expression 'swim gracefully' is not redundant, but the expression 'imagine interestingly' is. If we want to express evaluations of how people imagine things, we avoid the word 'imagine' and apply our evaluative adjectives either to some more neutral word, like 'depict' or 'describe,' or to the imaginer himself: we say 'he gave a delightful *account* of how it might be,' or '*he* was most amusing when asked to imagine a lunch with the prime minister.'

ii However, in general when x is said to be 'part of the concept' of c, we would not call anything c if it is not x; and yet one is not required to imagine everything amusingly, or interestingly, or even perceptively. One may, but need not give an amusing account of a death scene, or a fascinating account of how boring something would be, or a perceptive account of two un-named people discussing whether pigs have wings. Unless, therefore, we can specify conditions under which imagining must be witty, we will not be in a position to say, as we perhaps should if it is part of the concept, that we could not call this or that imagining, because it is not witty.

I do not think that in fact we can specify any such conditions. We could not say that whether an imaginative representation should be (for example) witty or not depends on the wishes of the person suggesting that one imagine. If he wants something witty, he can of course say so, and one may try to please him; but if he does not say, and what one produces is interesting or perceptive but not funny, one has done a perfectly good job of imagining.

The only other basis for specifying these conditions that is plausible is to say that the nature of the project determines the qualities required: that perhaps wit is inappropriate in an account of a death scene, and that perceptiveness is required in an account of an incident involving designated individuals or types of individuals (for instance, a stuffy professor and a sulky student).

Clearly, however, although we are not required to be amusing in imagining a death scene, we may do so, and not always unsuccessfully; and equally clearly, we can do a successful job of imagining an incident involving designated individuals without our performance being perceptive. I may produce a delightful account of a conversation between the prime minister and the minister of finance about the budget, employing no more intimate knowledge of these persons than is implied by the positions they hold.

One might in desperation suggest that certain qualities are called for *prima facie* by the nature of this or that imagining project, but that a performance may fail to have a property *prima facie* required and still be imagining, provided it has some other imaginative property. An account of a conversation between two named people may be unperceptive as long as it is amusing or beguiling or ..., but if it has none of these properties, one has failed to imagine.

Here surely we would be attempting to be altogether too solemn about imagining. It is not as if anything ever hinged on whether what a person had done was just exactly *imagining* something or not. We have no occasion to say, as if a ridiculous blunder had been made, 'I asked you to imagine it, and you have merely described it,' or 'That is not imagining: it is utterly unperceptive.' We may, if what someone offers when asked to imagine something is of an obvious and routine sort, go away muttering that he is an unimaginative fellow; but we would not know what to say as to whether his performance was imagining all right, but not very good, or not imagining at all. Never having had to make this decision, we do not know how to make it.

There is, however, at least one kind of occasion when we withold the designation 'imagining' from something on the ground that it is not good enough: we sometimes say 'I can't imagine it' when, although we have some ideas as to how something might be imagined, none of them seem very interesting. This may be enough to show that, whether in a formal way or not, evaluative notions are part of the concept of imagining.

It is relevant also that the adjectives 'imaginative' and 'unimaginative' do not express a quantity of something so much as a quality. The unimaginative person is not so much the person who does nothing by way of imagining something, as the person who does something but does it extremely badly.

A suggestion that someone imagine something is not, as the idea of formal conditions implies, a requirement that he come up to some standards, but an invitation to self-expression. We as it were point a person in a certain direction, and leave it up to him what he does next. It is as if, having shown a person samples of various different styles of painting, we set him up with canvas, paint, and a title for his work, and left the rest to him.

The foregoing account of imagining may, when set beside the precision sought in the Socratic tradition of defining words, seem outrageously rough. I have, for example, offered various lists of the qualities of an imaginative representation, and have sometimes written 'and' between the items on the lists, and sometimes 'or.' If this seems to anyone to be an objection, my reply is that I am not attempting the same enterprise as Socrates. I am not aspiring to succeed where he and others failed, I am not trying to specify what imagining is, but to supply reminders of what it is. Since we all understand the word, there would be no need to do this, were it not that we so easily fall into thinking of imagining as being something that it is not. Recognizing that saying how we imagine something is not itself imagining it, we urgently ask: 'Then what is the imagining itself?' and since the expression 'saying how we imagine' suggests a report, we suppose that what we say is a description of something; but since all that is going on overtly is the imaginative representation we give, the imagining itself, we think, must be going on or have gone on in our minds. In this and similar ways we are enticed away from what we all know about the use of the word 'imagine,' and come to think of it as a hidden thing, about which other people have to make guesses and surmises; and, moreover, as something not beguiling by nature, but (as one might put it) by force of circumstance: if imagining is a mental process occurring in idle moments, then since it is private and not directed to anything, it is likely to be free-ranging and self-expressive. But that, on this line of thinking, is a contingent fact about it, and idle mental processes are not less imagining if they are uninteresting and random.

People who fancy themselves hard-headed may reject the idea that imagining is a mental process, and say that it is right out there in the open: it is the process of describing a scene or an incident, or of enacting it, or of making a cartoon drawing of it. But (i) it is not for example the *activity* of cartoon-drawing, but the cartoon that we draw, that may be how we imagine something, and (ii) we may draw a cartoon or enact an incident, but reject it as not being how we imagine something, and therefore it is not whether I have produced a representation, but whether it is successful in expressing how I imagine something, that is the important thing. If imagining is thus neither a mental nor a physical process, and is not what is produced by the physical process, it may seem utterly bewildering, what it could be.

It is only bewildering, however, as long as we suppose that imagining must be some kind of a *phenomenon*, something that happens or exists, whether in the inner or the outer world. The problem is solved, I suggest, when we remember the way in which such evaluative notions as being interesting, perceptive, or amusing are part of the concept of imagining. In saying how we imagine something, we are not reporting on, describing, or revealing any state of affairs, but declaring the representation we then offer to be up to a kind of standard: to be an adequate expression of our taste, wit, and perceptiveness. In this way the account of imagining that has been sketched cuts us completely free of the dreary milieu of mentalism *v* behaviourism. If our account is rough, its roughness is not such as to interfere with its performance of this philosophical service; and the philosophical service of such a theory exhausts our demands upon it.

Let me conclude by saying something about the species of direct-object imagining that involves no development, the cases where the object of the verb is a proper name or a definite or indefinite description. Let us call this 'static' imagining.

While what I called extended prescription direct-object imagining is, as I hope I have shown, fully expressed in descriptions, play-acting, or cartoon drawing, static imagining is most readily conceived as being a matter of generating mental images – partly I think because if when asked to imagine my grandfather I *describe* him, there seems no distinction between imagining him and simply remembering his appearance until I assure you that I see him clearly in my mind's eye and that my description was read off from this image, and partly because there seems really nothing else to *do* in the case of static imagining than conjure up a picture. The absence of any requested development leaves no place for wit or perceptiveness.

Is there then at least this one kind of imagining which is along the lines we originally supposed all imagining to be, namely, conjuring up a mental picture or image? I am not unalterably opposed to saying so, but I will point out certain difficulties in the notion.

In the first place we do not at all naturally use the word 'imagine' in quite that way. If someone asks me to imagine my grandfather, I am inclined to wait for the rest of his request; wait to be told what I am to imagine him doing, or to what state of affairs I am to imagine him reacting. And if upon expressing this expectancy the other person replies 'No, just imagine him,' I would be likely to say 'Oh, you mean conjure up a mental picture of him.' The word 'imagine' by itself does not seem to specify this, and we require some further explanation if this interpretation is to be put on its use.

Secondly, even given this interpretation, it seems extremely unclear when one should say that one has accomplished this task. If I remember my grandfather as a short man with a grey beard and a watch chain strung across his vest who was fond of reading in front of a fireplace, and conjure up a picture with those ingredients and the rest sketched in vaguely, have

I imagined him? Or if I vividly see now his eyes, now his smile, now his hair line, now his posture, until I have surveyed most of his characteristic features, but can't see them all together? Or if I can only picture him when angry, although that was not a very characteristic state of his, and I was not *asked* to picture him in any particular state?

And thirdly, while the kind of direct-object imagining we have mostly discussed has many functions in our lives (it entertains, helps us to understand, displays our personality), there seems extraordinarily little point in 'static' imagining. When you ask me whether I can imagine University College, all I can do is assure you that I can or that I can't. And while this information may be of some interest to you, it is an abrupt conversation-stopper; while the general function of the word 'imagine' has just the opposite tendency.

Some questions about dreaming

1 Is it possible that we do not experience anything when we dream, but that something happens in our brains when we sleep, and because of this, when we awake it seems to us that we have had the experiences that we call dreams?

The above possibility is, of course, counter-intuitive, because we have the very strongest conviction that in dreams we do have experiences, broadly resembling waking experiences, of people and places, of conversations, of doing things and having things happen to us. Yet is it not possible that we are not in fact conscious of these things at the time the dream is supposed to have occurred, but that what then happened was a certain activity of our nervous system such as, had we been conscious, would have resulted in our having the experiences that it seems to us on waking we had? According to this supposition, it is not the case that *nothing* happened as we slept, but that what did happen was not what later seems to us to have happened, but rather the neurological correlate that such experiences would have had, had they been conscious; and that in remembering or seeming to remember dreams, the same sort of neurological mechanism is at work that enables us to remember waking experiences: the neurological correlate leaves traces of some kind which later in some way are responsible for our recollections; and as long as the neurological correlate occurs, it is not necessary to our remembering (or seeming to remember) that there should have been any experience of which it was the correlate.

Reprinted from *Mind*, LXXX, NS 317, 1971

This conjecture is stated in very general terms, and includes *no* sophistication as to how exactly the nervous system operates in these regards, but this should not be an objection. It is extremely unlikely to make any difference to the discussion we are embarking on, whether the nervous system works this way or that. *A priori* one would think that the only discoveries in this area that would bear on our discussion would be either (a) that we remember by 'storing' the shapes, colours, sounds that we experience in just the form in which they are experienced, or (b) that the nervous system is not involved in memory, but that (perhaps) spiritual substance does the job quite independently of our bodies. But given only that the nervous system is indispensable, our question does arise.

A thesis of the above kind would seem to square well enough with various reasonable assumptions, if not known facts:

1 It seems reasonable to suppose that there is some correlate in our nervous system of every experience we have, that this correlate is causally responsible for our experiences, and that the mechanism by which we remember an experience is not one of somehow storing the experience itself, but of somehow storing the neurological correlate of it.

2 It seems reasonable to suppose that if the neurological correlate of an experience occurs, it will not be necessary to the functioning of the recollective machinery that the phenomenological correlate should have occurred, just as, if electrical impulses in an amplifier can cause the impression of someone singing, then although normally a singer is necessary to provide the input, it will be theoretically possible to deliver the output without the aid of a vocalist.

3 It is reasonable to believe that one has to be conscious in order to experience anything, and that since in sleep one is not conscious, no experience occurs: there is no phenomenological correlate.

4 One is easily able, on this supposition, to account for the fact that we seem to *recall* dreams: something (namely the neurological correlate) did occur while we slept; and it is by virtue of the functioning of the same mechanisms as normally operate in remembering that we have the impression that we dreamt.

5 One is easily able to account for the fact that dreams are at least roughly dateable by the occurrence during sleep of such behaviour as smiling, frowning, muttering: something did happen at that juncture, and it is quite believable that we should acquire a disposition to react directly to the neurological correlate of an experience, in the same way that we would have reacted to the experience itself.

6 One might with this thesis also be able to explain some bizarre features of dreams. A young man dreams of a delightful flirtatious conversation with a girl who, however, looks like Joseph Stalin. These things, one feels, just will not go together into a possible experience, but it is quite possible that the neurological correlates of the two things should occur at the same time.

In spite of these considerations, I find that I cannot believe this theory about dreams. I suspect that any reasons I give may be only a justification of what I (do not) believe on instinct, but I do offer the following counter-arguments:

1 It is somewhat surprising, if this theory is true, that we do not remember things that happen around us when we sleep, such as conversations that occur in our neighbourhood. It is at least possible that when a sound is made near us, the neurological correlate of our hearing a sound occurs, but just does not, as one might put it, 'reach consciousness'; and if it does occur, one would expect on this theory that it would sometimes be remembered upon waking. It is, of course, an empirical question just where the neurological activity stops, and if it were found to stop just at the eardrums, then this point would turn out to be worthless. Yet not only might it not stop there, there is a presumption that it does not in the fact that teaching can be done with microphones under our pillows while we sleep. And this phenomenon seems to count in another way as evidence against the thesis: people *absorb* what is fed to them during sleep, but it does not appear to them later that they have *heard* what they thereby acquire, the way it seems to us later that we have experienced the content of a dream. They do not know *where* it came from.

2 It seems probable *a priori* that there would be some neurological feedback from our actually being conscious of something, so that there would be different traces (or whatever) from the same neurological process when it did and when it did not have a phenomenological counterpart, and that this difference would show in the way it appeared to us when we remembered it. (This might explain why it is that when a person has been fed information while asleep, though he absorbs it, he has no impression as to how he acquired it.)

3 In the cases where one can be persuaded that something one might otherwise have been inclined to suppose to be an experience is not in fact – in the case of meaning what one says, for example, or of intending – it is not simply the *arguments* that are convincing, but the arguments clear away the prejudice that prevents one from seeing that in fact that is just

how one remembers many cases of meaning something or intending something. You find that after all you *remember* no experience in these cases (or no experience that you would care to call 'meaning' or 'intending'). But no matter how strongly I am tempted by the arguments supporting the above thesis about dreams, I cannot rid myself of the impression that I do remember dreams as experiences.

4 We react to dreams: they are delightful, disturbing, terrifying. And they are that way, not merely now when we remember them (perhaps hardly at all when we remember them), but at the time: people smile as they dream, or go rigid with fright. But while it is natural and typical of people to react in these ways to such experiences as chatting with a delightful person or being chased by tigers, it is not so comprehensible that they should so react to the neurological correlates of these experiences. People *could* no doubt so develop that the occurrence of the neurological correlate gave rise *directly* to the reaction. Even if it is tigers themselves that are terrifying and conversations themselves that are delightful, and we originally react to *them*, those reactions could set up a secondary pattern of such a kind that whenever the nervous system is in the state it is in when we perceive a tiger, it moves directly to the state it is in when we are terrified. But this is surely an enormous hypothesis, in view of the tremendous complexity of the nervous system, the many different conditions of it that would be involved in seeing tigers of different sizes, shapes, and postures, and the slight neurological difference between seeing a real and seeing a paper tiger, which does not terrify.

Moreover, if a secondary pattern of this kind were set up there would be a problem as to whether it would no longer be the case that we were in the ordinary way afraid of tigers, or whether we would be *doubly* afraid when we saw a tiger, first because the beast was there, and second because our nervous system was in one of those states that leads directly to fear; and as to whether we could distinguish between the fear inspired by the tiger and that begotten by the condition of the nervous system. Would the latter be relieved entirely by turning our back on the beast so that the nervous system was no longer in one of its fear-begetting conditions?

5 It seems to be the case that we can remember dreams in words, that is, that the things we say about a dream can be our *first* recollection of it. And it is comparatively easy to understand our being able to do this if we suppose that we did have experiences such as we upon waking describe. After all, we have learned language partly as a way of talking about things of just

that general kind: people, tigers, long avenues of trees; so that the word 'tiger' now comes immediately to the lips when we see a tiger. (We do not have to *judge* it to be a tiger, a feat some people think we could not do while asleep.) But we have not learned to recognize neurological conditions as the counterparts of being chased by tigers, and if we had, it would be odd that we should report the occurrence of these conditions as 'seeing a tiger,' rather than as what they are: being in the neurological condition appropriate to seeing a tiger.

These counter-arguments, I believe, will account for all but one of the considerations advanced in support of the thesis that dreams only seem to be experiences. In general, those considerations were intended to show that the known facts or reasonable suppositions about dreaming could be adequately accounted for in terms of neurological correlates, and that even if the common sense view of dreaming also accounted for these facts and suppositions, the neurological correlate thesis was preferable because it did not require us to say that we could have experiences when we are not conscious. The counter-argument so far makes it moderately clear that the neurological correlate thesis does *not* account for the known facts or reasonable suppositions, while the common sense view does; but there still remains the final and supposedly crucial consideration that the common sense view requires us to say that it is possible to have experiences when we are not conscious. How much of a difficulty is this?

I suggest it is no difficulty at all. It is of course true that we can not smell, see, hear, or even imagine things when we are not conscious; and we might carelessly generalize from this to the conclusion that we can experience *nothing* when not conscious. But surely if we asked any careful person whether this generalization would hold, a moment's reflection would suggest the case of dreaming as a strong and obvious counter-instance. At that intuitive level of argument, the case of dreaming is just as good evidence against the thesis as the cases of seeing or hearing are in support of it. And as far as I can see there will be no considerations of a more abstract kind that bear on the question. If not being conscious is just not being accessible to external stimuli, then *of course* we will not be able to see or hear; but if dreaming does not require external stimuli, then there will be no conceptual impossibility about having dream experiences while not conscious. (Perhaps if we think of being conscious as a matter of having the lights on on the internal stage, that being as it were the condition on which stage persons or props of any kind will be perceptible, then of course dream persons and props will not be perceptible *either*, when 'the lights are off.' But that is just a roundabout way of making the generalization that we concluded was unwarranted. If we must entertain such a picture, we will simply have to amend it in such a way as to allow for the possibility of dream experiences while 'the lights are off'; for example, by saying that dream tigers and forests are self-illuminating.)

II *Does the concept of error apply to our recollection of dreams?*

There are two distinct claims that one might be inclined to make as to the application of the concept of error to dreams: (1) that in the case of dreams, whenever we really seem to remember something, then something of that kind did in fact occur in our dream: seeming to remember is infallible; and (2) that we cannot either get it right or get it wrong; it simply seems to us that something happened, and that is the end of it. What is interesting is not (cannot be) whether we got it right, but only that it should so seem to us. Let me first make a few general observations about these possible claims, and then discuss each of them in turn.

Clearly, the issue we have just been discussing will have important bearings on either claim: it would, for example, be very difficult to see how a dream recollection could be *right* unless it was in line with what had actually happened, and therefore an infallibility thesis would seem to require that we do experience dreams, although of course the converse does not hold. The thesis that we do experience dreams would in no way entail that our recollections of them could not be in error. (There might, however, be a strange sense in which we could get it right if the neurological correlate of what we seemed to remember occurring did occur. But even under these auspices, and even in some future time when neurological correlates, if there are such things, may be mappable, for practical purposes saying that a person was not wrong about his dream recollections would only be tantamount to saying that there is a special kind of *current* human fancy that is not imagining or whimsy or pretence but – what would we say? – comes from deep down?)

The thesis that dream recollections can be neither right nor wrong would, on the other hand, at least be strongly implied by the thesis that dreams are not experienced at the time they are thought to have occurred. It is at least not obvious, however, whether the converse of this holds. One would certainly be much inclined to say that if a dream experience of a certain description did occur, then whether or not we ever *know* it, our dream recollections are in fact right or wrong, accurate or inaccurate. But there is, perhaps, an equally strong inclination to say that if there is absolutely no way of knowing whether our dream recollections are accurate, then it makes no sense to claim accuracy for them, or inaccuracy either.

It is extremely easy to confuse the two theses we are now considering, since one way of expressing the view that the concept of error does not apply is to say that one's dream recollections can not be wrong; but from this it appears to follow that they are necessarily right; and that is another way of saying that they are infallible. But what one *means* in saying that they can not be wrong is not that they are always right, but that it makes no sense to say that they are wrong, or right either. We can see the absurdity of the inference to infallibility very clearly if instead of saying that dream recollections can not be wrong, we say that they can not be right. This is an equally good way of expressing the thesis that the concept of error does not apply; but proceeding in the same way that we proceeded from the 'can not be wrong' way of expressing it, we would arrive at the conclusion that dream recollections are necessarily erroneous.

Let me now make some brief remarks about the infallibility thesis, and then at greater length consider the view that dream recollections can be neither right nor wrong.

1   *Are our memories of dreams infallible?*
There seems to be very little to recommend this supposition, except perhaps as a disguised form of its sophisticated cousin – the thesis that dream recollections can be neither right nor wrong. We would *treat* dream recollections the same way on either theory, that is, we would not ask whether they were correct but would treat the fact that a person does believe he had a certain dream as the important thing; and the difference would be that our *reason* for not enquiring whether the dream occurred would in one case be that it made no sense, while in the other we did not or would not doubt that it did occur.

This refusal to doubt could not, however, be justified in any way, and could only be treated as an article of faith. If one were struck by the fact that we have no other means of access to dreams than our present recollections of them, and if one were worried about the possibility that we might be radically wrong, either about *what* happened or about whether anything happened, and would have no way of detecting such error, one might wish to lay it down that we simply do not make mistakes about dreams, that whatever really seems to us to have happened did happen.

One could not, however, say that such a principle was a *discovery*, based, for example, on the fact that it had never been found to fail that a person actually had dreamt what it seemed to him that he had, or on the fact that the special brain cells used for remembering dreams are not liable to malfunctions the way cells used for other kinds of remembering are. Either of these grounds for the infallibility thesis would have to rest on just what we are supposed not to have: access to what happens in dreams independent of what it now seems to us that we dreamt. Even if we found that there were special brain cells used in remembering dreams and that they were peculiarly well formed, we would still need to test their output against their input to determine whether their admirable form was such as to deliver recollections unerringly.

2 *Does it make sense to wonder whether a person has remembered a dream correctly?*

Since to do otherwise would, monumental sophistication aside, settle the question without more ado, in this part of the discussion I shall assume that dream experiences do occur during sleep, and simply ask whether, for whatever reasons, we must ignore this fact in dealing with dream recollections.

I shall also assume that we are able easily enough or often enough to distinguish cases of story-telling or spoofing from genuine cases of dream recollection. But we will see later that there is a certain problem about this.

One way of making out a case for the view that it makes *no* sense is to draw up a comparison between ordinary remembering and the recollection of dreams, with a view to showing that in the latter case we have none of the aids to remembering, or checks on remembering, that are available in the former case, and that therefore there can be no substance to a claim that we have got it right about a dream.

In the first place it is obvious that in the case of dreams there is available none of such supporting evidence as photographs, letters, maps, the testimony of other people as is available in support of other claims to remember.

Secondly, although the amount of such evidence varies in individual cases down to nil, there is another important type of consideration available for ordinary remembering and not for dreams, namely what we might call relational clues: I remember that he spoke of his children's problems but can not right away remember what he said, but I can remind myself by reviewing what I know of their problems or by finding out what their problems are; I remember that someone made a certain remark but can not remember whether it was Mr A or Mr B, but if I know or find out that Mr A is too witless to say such a thing while Mr B is not, I may have some confidence that it was the latter; I seem to remember a train of events in a certain sequence, but on the other hand I know that it is extremely unlikely that those events should occur in that sequence, while quite likely that they should occur in a different order, and I conclude that that is how it must have been; or a conversation as I first remember it would not make sense or would not make the kind of sense I remember it did make, but with certain revisions *would* make that kind of sense. These clues are not available in the case of dreams. A man, even if he is identified as someone one knows, will not necessarily or even likely speak realistically about his children's problems in a dream, and may speak of his children's problems when (in real life) he has none; persons encountered in dreams do not generally have sufficiently well-defined personalities to enable one to conclude anything as to what they would or would not say; there are few if any limits or even faint probabilities as to the sequence in which dream events will occur; dream conversations do not have to make sense or to make nonsense either. And so on, I think, for any other sort of relational clues one might suggest.

Thirdly, with ordinary remembering one can sometimes remember more fully or more exactly just by intensity of concentration. One perhaps tries to think of absolutely nothing else, and then suddenly it comes clear. But this process brings it about that we get it right only if, for example, another person then says 'Yes of course! How stupid of us not to remember that!' or if our spectacles do turn out to be just where we suddenly remembered leaving them, and so on. But in the case of dreams there is never this grounding of the recollections that in this way seem to come clear, and therefore there is *no difference* between getting it right and its merely appearing differently to us after having tried hard to remember.

This comparison might be taken to show either one of two things: that remembering dreams is so radically unlike ordinary remembering that none of the things on the basis of which the concept of error applies to ordinary remembering hold in the case of dreams, and therefore the concept of error cannot apply; or merely that remembering dreams is very different from ordinary remembering, and therefore questions of truth or accuracy are ever so much more murky and uncertain in the case of dreams. I wish to defend the latter view. To keep my arguments separate I will number them.

i We said that in the case of ordinary remembering but not in the case of dreams there was available to us such evidence as photographs, documents, the testimony of other people. And this is generally true; but it is not true either that we always take advantage of such evidence, or that in every case there is any of it available. And one does not have to reach for weird or out of the way cases where there is no such evidence available: there is no possible evidence of the pacing around I have done this afternoon in my room, of the rabbit I saw from my window this morning dancing the Charleston, or of any number of the things that we see or do in the course of a day; yet we remember many of them, and have no doubt that they occurred.

To take a more classic philosophical case, let us consider a conversation at which no one was present but myself and my dying friend. We say he was dying in order to eliminate any later testimony of his as to the course of the conversation; and to eliminate 'relational clues' we will suppose that he was somewhat delirious, did not reminisce about his past or mention his family or friends, spoke in a way very uncharacteristic of his normal personality, and directed me to a buried treasure which on no interpretation of his instructions was discoverable. On the principles governing the above claim about the recollection of dreams, I think one would have to say that it would make no sense to suppose that I remembered this conversation correctly; yet while I might have some difficulty remembering some parts of such a conversation, and while no one could have any grounds for *affirming* that I got any of it right, it is monstrous to say that the question whether I got it right makes no sense – if this implies, as I think it does, that 'I don't know whether to believe him or not.' 'I believe him' and 'I don't believe him' are equally absurd reactions to my account of the conversation.

To this someone *might* reply that although in a case like this there is in fact no available evidence, still it is the *kind* of case in which there might have been witnesses, tape recordings, or relational clues. We know what it would be like to verify it, even if we cannot in fact do so. But I am not clear what difference this fact would make, if it were a fact; nor do I know what to make of the idea of someone else being present at a conversation at which no one else was present.

In this connection the following consideration may not be altogether fanciful: it might be suggested that we *train* our memories, and that they *become* reliable, through repeated experiences of being corrected about our recollections, and through the establishment of the habit of connecting the thing to be remembered in various ways with other things in the world. The result of this training, it might be suggested, is a certain presumption of reliability, even in cases, such as the death-bed scene, where there is little or no possibility of corroboration. But, this line of argument might proceed, no such training is possible in the case of remembering dreams, both because our recollections of them never stand to be corrected, and because there is no way of connecting them with other things in our world. Hence, while it is acceptable to say that in some cases of ordinary remembering we simply remember, it is not acceptable in the case of remembering dreams, because there can be no established presumption of reliability in this case.

Surely, however, this is an argument that can cut both ways. If the reliability of my memory for conversations has been built up through conversations with my wife, can this reliability carry over into the case of conversations with strangers? If it cannot, then there would be very little general presumption of reliability of anyone's memory, since so many of the matters we are called upon to remember are not matters of which we have long experience; while if it does, then it is difficult to see why it should not equally carry over to one's recollection of dreams. It is not after all as if dreams were radically unlike other experiences. The constituents of dreams – people, buildings, animals – though they may be related to one another in weird ways, are generally speaking all items of which we do have waking experience, and are just the kind of things concerning which, on this supposition, our memory becomes reliable.

ii The claim that the concept of error does not apply would be a good deal more believable if we told of our dreams with an untroubled mind, so to speak: spontaneously, easily, and with no sense of trying to remember, despairing of remembering, wondering whether we have missed something, no sudden realization that we have been omitting an important feature of the dream. But that is not how it is. We try to remember dreams in just the frame of mind that we try to remember anything else, *regarding* the concept of error as applying, although unable to use most of the tactics and devices that are useful in other kinds of remembering.

iii It is of some importance that those tactics that I called relational clues are not such as to show us what happened, but only such as to help us *remember*: if I remember that he spoke of his children's problems, but can not remember what he said, then a review of those problems will not show me what he said, but only put me on the road to *remembering* what he said; and it may turn out that what at last I remember is utterly surprising in view of what I know about those children. As far as the relational clues are concerned, the recollection has to be self-supporting, and to this extent is in no stronger position than the recollection of dreams.

iv Although there is not evidence of the alternative-record kind (photographs, recordings, the testimony of other witnesses) as to what a person dreamt, it is not the case that we can never be certain what a person dreamt. Consider these cases: (a) There is an automobile accident, and when the news of it first breaks my friend is visibly shaken by it, because he says he dreamt just last night that there would be an accident at that place, involving those persons. He goes on to describe some of the details of the dream; and upon visiting the accident scene is utterly appalled to find it all to be just as he had dreamt. In such a case we would surely be as certain as we are of most things that he did have that dream, especially if in general we knew him not to be a practical joker or a person with a macabre imagination. (b) If I had a drug which, within a short time after a person took it, during which time he did not sleep, made it seem to him that he had had a certain dream last night, I would be as certain as I am of anything that he did not have that dream, and that it only seemed to him that he did.

v We might in general be inclined to say that all we are certain of is that it really does seem to a person that he has dreamt: as to the dream itself we cannot ask. But (a) the above two cases seem to count against this view. In the case of the man who dreamt of the automobile accident, what everything convinces us of is not that it really seemed to him that he had that dream, but that he did have it. The drug case is even clearer: we are at the same time convinced that it really did seem to him that he dreamt, and that he did not dream. And (b) we perhaps take it too much for granted that its seeming to a person that he dreamt is the clear notion, and his having dreamt is the dark one. The seeming, we may feel, is something present and examinable, it is right here in our lab, while the dreaming is not only now lost in the mists of the past, it would not be accessible even if it were now present.

But it may appear just the other way around if we ask ourselves under what conditions outside philosophy one would *say* that it really seemed to a person that he had dreamt. We would say it when we had reason to doubt that he did dream, or knew for a fact that he did not, and therefore the cases would be rare in which there was a justification for saying it. In the ordinary case, what we become convinced of is not that it seems to a person that he dreamt, but that he did dream.

Here one might of course be inclined to reply that while we regard ourselves as, and talk as if we were, reaching conclusions as to what people dreamt, or whether they dreamt, still the evidence we have is all evidence as to whether it seems to them that they dreamt, and therefore that is what we are really concluding; but since it is not useful in practice to make any distinction between this and the dreaming itself, we say that we become convinced that he *did* dream.

But while much of the evidence is of that kind, it is not peculiar to dream recollections that this should be so, but is typical of all remembering. Moreover, some of the evidence has to do with the dream itself. It is, for example, evidence against a man's having had a terrifying dream if he was observed to sleep peacefully during all the time when he regards himself as having so dreamt; and in our drug example, the fact that he did not sleep at any appropriate time is even stronger evidence. But since we seldom have occasion to mount a watch on a person while he sleeps, the following case may be more interesting: a young man tells a dream to a young woman he admires, which is suggestive, charming, flattering, but somehow not believable as a dream. We think that either dreams do not run that way or it would be a fantastic coincidence if he had happened to have a dream that served his romantic purposes so admirably. We therefore do not believe that he so dreamt, *and from this conclude that it does not really seem to him that he did.* We could strengthen the example by supposing that he was a very good actor, and went through all the agonies of trying to remember, sketching it dimly at first, filling in details, making revisions, pausing in rapt concentration, experiencing sudden flashes of recollection, so that all the evidence would point to its really seeming to him that he had dreamt. But in spite of this we conclude that it does not so seem, on the grounds that he did not so dream.

III *Can dream recollections involving the identity of persons, places, buildings, etc. be correct or incorrect?*

As an alternative to the very general thesis we have just considered regarding the applicability of the concept of error to dreams, one might put forward a number of special theses, one of which we will now consider and another of which will be examined later on. Whether these special theses might cumulatively be equivalent to or support the general thesis I will not consider.

If I say I dreamt that I had a conversation with Ryle, or that University College was on fire, it seems incontestable that it makes no sense to ask whether it was really Ryle, or really University College. A person would be making an ass of himself if he said it could not be Ryle, because Ryle is in Oxford and I am in California, or because he had asked Ryle, and he had no recollection of it; and it would be equally foolish, if one knew Ryle to be travelling in California or to be egregiously absent-minded about conversations, to say that it might have been he. Does this show that the concept of error does not apply to recollections involving identity?

I think it does not: it only brings out something about the *concept* of dreaming. I said I *dreamt* it, and this *means* that no real conversation occurred, that it would be senseless to see if Ryle himself would confirm its occurrence, or to enquire whether he and I were geographically close enough for it to be possible. It does not touch the question of whether I might have misremembered my dream: of whether (1) the person with whom I chatted looked or talked like Ryle, or (2) of whether, regardless of what he looked like or said, in my dream I *took* him to be Ryle.

1 *Can one misremember the appearance of (for example) a person in a dream?* This I think is just a particular case of the general question discussed in Section II, and it should be moderately clear now, (a) that if we do not initially remember the appearance of someone of whom we have dreamt, it makes sense to try to remember it. We will see later that we can not assume that a person in a dream *has* any appearance; but this I think only adds a further dimension to what we may try to remember. We may try to remember whether the person we chatted with 'had any appearance,' and if so, what? And (b) we may initially remember something one way, but then have it strike us that we were wrong, and that the dream actually was quite different in various ways. I remember that I had a conversation in a dream with Ryle, and initially I am inclined to say that he looked like Ryle; but then it strikes me that while I *took* the person I spoke with to be Ryle, it was in fact a large middle-aged woman who sat nursing a child as we talked. That is possible; and hence it is possible that, conscious of the false assumptions I may make as to the contents of my dreams, I should very often ask myself whether I have got it right.

Perhaps, however, the person who claims that it makes no sense to suppose that we may have misremembered a dream is less interested in this stage of the proceedings than he is in the point at which it does seem to us that it was in fact (for example) a middle-aged woman nursing a baby. He might say that when we remember that we had a conversation with Ryle, it is not so much that we *remember* that the person *looked* like Ryle as that we *assume* he did; that the only *remembering* as to the appearance of the person comes when it strikes us that it was in fact a middle-aged woman; and that at that point we can not ask whether we might be wrong.

If this distinction between assuming and remembering is essential to the claim that we cannot misremember what we ultimately seem to remember, I think it would be very difficult to make it stick. It of course seems very clear in the case in which it suddenly dawns upon us that we have been wrong: *then* we perhaps see clearly that we were just assuming that the person resembled Ryle in appearance. But in the ordinary case in which there is no dawning of new realizations we could not distinguish between assuming and remembering. Assuming does not have a phenomenological character such that by careful examination of ourselves when we are assuming we can discover that that is what we are doing. It only takes on a character (of foolishness?) *after* we have discovered that we were merely assuming something.

'Really remembering' similarly seems to have a special character only in the case in which it dawns on us that we have been making a mistake: there is a clarity and excitement about it that is not otherwise present. But not all cases of remembering are of the sudden dawning kind, and there are no special marks of remembering in the common run of cases.

Moreover, it does not seem impossible that a sudden dawning should in turn be subject to correction. It strikes me that the person I took in my dreams to be Ryle in fact bore a resemblance to Miss Anscombe, but wait: perhaps that was not the same dream. I had other dreams last night, in at least one of which I remember Miss Anscombe figuring; and I also know myself to relish the contemplation of the very strange things that can occur in dreams. Perhaps the combination of these factors resulted in its falsely seeming to me that I took a middle-aged woman to be Ryle.

Here the fact that one may have almost no resources with which to answer such questions may become the key consideration. But I do not think that this should deter us from saying that the questions make sense. In the first place, while we can not have a *high* degree of confidence as to the answers, we are not always absolutely without resources. I may in a case like the above remember quite a few of the details of the *other* dream in which an Anscombe-like figure appeared, and may also remember an Anscombe-like figure saying things which I do not remember as being part of that episode but which could be part of the conversation-with-Ryle episode. And the supposition that there were two Anscombe-episodes may be reinforced by a recollection of having awoken briefly and mused on the fact that I was having so many dreams involving an Anscombe-like figure. Or when I put to myself the question whether I am merely driven by my fascination with the weirdness of dreams to say that *this* dream had this weird feature, it may seem to me in this case simply untrue. None of these is of course a weighty consideration, but neither are they worthless.

Secondly, I think we are too much inclined to take a case like this in which doubts abound as the test case, and to judge every case of dream recollection by it. One certainly can find cases in which there is nothing certain or even moderately probable. But that does not show that doubts abound in all cases. One can raise the same doubts as to waking experiences, and once raised they will not likely be resolved. But in many cases we would treat these doubts as being academic, and simply would not in fact doubt. (Consider, for example, the case mentioned earlier of the man who dreamt of the automobile accident.)

2 *Can we misremember whom we took a person in a dream to be?*
There are several quite persuasive considerations that might be advanced in support of an affirmative answer to this question:

i In a dream it is not on the basis of a person's appearance that we take him to be so-and-so. We do not size him up and judge him to be so-and-so, nor do we know right off that he is so-and-so, but one's regarding him as being so-and-so and the figure he cuts in the dream are as it were independent constitutents. I can take a person to be Ryle although he looks like Miss Anscombe, and although it would be quite impossible for me to behold an Anscombe-like figure and say 'I know him (her?), he's (she's?) Ryle.' And in the same way it is not on the basis of what people say or do in dreams that we take them to be a certain person. A dream figure could be talking like Hegel or dancing the Watusi and still be taken by me to be Ryle. Therefore, a large body of possible evidence as to whom a dream character was taken to be fails to operate in dreams the way it might in waking experience.

ii This point is further reinforced by the fact that a dream person need not have *any* characteristics: dreams are sometimes like beforehand rehearsals of conversations we are going to have, for example, with a prospective employer, where we do not always imagine any face or figure playing the part of the other person. We just suppose that there is someone there, and do not fuss about what he looks or sounds like.

iii Our taking a dream person to be so-and-so need not itself have been an event or series of events in the dream. We need not have thought 'It's Ryle,' or addressed him by name or referred to him by name in speaking to any other person figuring in the dream. Our taking him to be Ryle may come to a head only in our saying upon waking that we did. There therefore seems not necessarily or even generally to be any dream *phenomenon* here to be either remembered or misremembered.

iv It is not absurd to regard dreaming as being at least importantly like imagining, that is, to think that in dreams we do something like illustrating a story with pictures, and that sometimes the wires get crossed so that we produce all the wrong illustrations, but that the same crossing of wires results in its seeming to us as we dream that we are illustrating appropriately. If I set out to imagine Iris Murdoch and produce a likeness of Elizabeth II, one can not say that I have not imagined Miss Murdoch, but only that I have done a very bad job of it. My setting our to imagine her makes it her that I imagine no matter how ineptly I do so. Similarly, it is not the case that I do not regard a dream person as being Ryle if the illustration of him that I produce is Anscombe-like. In this way *regardings* are not falsifiable.

I described these considerations as very persuasive, and I am by no means certain that the conclusion they are supposed to support is erroneous. But I will offer one or two points that I think might at least lead one to doubt it.

i The fact that the person taken to be so-and-so *need not* resemble that person either in appearance or in behaviour does not seem to me to show that it can never be a relevant consideration whether there is such a resemblance. Might this not depend among other things on the dreamer? If he was very much given to having the mixed-up sort of dreams we have been describing, then it would be no evidence at all; but if his dreams ran heavily to being coherent in this regard, then the fact that a particular dream had this kind of strangeness might justifiably lead him to doubt whether he had taken an Anscombe-like person to be Ryle. And such a doubt need not itself be unsettleable. Perhaps on further reflection he will recall that he in fact had two dreams, and will be able to spell out to himself in some way how he has run them together.

ii The fact that taking someone to be so-and-so need not at the time of taking consist of anything, that is, need not be an action or an event, and in that way a normal object of remembering or misremembering does not seem conclusive as to whether it is possible to misremember it. It was after all *at the time* that I took him to be so-and-so. If I say I had a dream about Ryle, I do not *now* judge from my recollection of his appearance or of the things he said that that was who it was. (Indeed were that all I had to go on I could not say that I had a dream about Ryle, but only about someone very Ryle-like. Dream persons, no matter how like real persons, can not *be* those persons.) I *remember* that I took him to be Ryle. Might I be wrong about this? Suppose that on thinking further about the dream I seem to remember someone saying Wisdom will be here soon, and then someone having arrived to whom I spoke, and that he is the one I now seem to remember as Ryle, and I do not remember any surprise that although Wisdom was expected, Ryle came, and further I am in general apt to become muddled about English philosophers, and upon considering these facts the conviction that I took him to be Ryle fades: might it not then be quite doubtful whether I had taken him to be Ryle? Of course, none of these considerations is weighty: the shift of identities is just the sort of thing that in dreams is taken to be a matter of course. And in such a case I could hardly become satisfied that I had in fact taken him to be Wisdom. But the point is that a reasonable doubt now exists as to whom I took him to be, and this is what is not supposed to be possible in the case of dream recollections.

IV *Does it make sense to try to remember more of a dream than one initially recalls?*

In the daytime world, if one is talking to someone, there is someone there; it is either man, woman, or child; if a woman, she is dressed in a certain way, is either pretty or plain, has either one facial expression or another at any given juncture in the conversation; she is sitting, standing, lying down, walking, running, or dancing; the conversation occurs either indoors or outdoors, and if indoors, in a room that is large or small, elegantly or plainly furnished, a sitting room or a bedroom. And while there may be many circumstances of a daytime conversation that were either not visible to a participant in it, or not noticed by him, we know *a priori* that there is an answer to such questions as whether it was man, woman, or child, whether the conversation occurred indoors or outdoors, and therefore it makes sense to urge a person to try to remember such particulars.

But it is different in the case of dreams, where we can have conversations with persons who are neither male nor female, neither blonde nor red-headed, neither formally nor casually dressed, neither standing nor sitting, neither indoors nor outdoors. We can even, without it striking us as in the least marvellous where the other voice comes from, have conversations in which the other person simply has no physical presence. It therefore makes no sense to say to the dreamer (or for the dreamer to say to himself) 'Come now. She must have been young or old, pretty or plain, seated or standing, and it must have occurred either in the daytime or at night, either indoors or outdoors. Do try to remember.' The person who says this is pressing a logical mistake on us. And it is even a mistake, I think, when one remembers in a dream having been charmed by a room or annoyed by a facial expression, to assume that it must have been by certain recollectable characteristics of the room that one was charmed, or by the actual occurrence of a facial expression in the dream that one was annoyed, and so try to remember it. It is possible in dreams to have these reactions without there being anything in the dream to which one is reacting.

And we do not need to agree that dreams are such *queer* things as this for this point to be valid: if we *are* in a room and there *is* a human figure there with whom we are chatting, but his face is averted when he makes some particular remark, it makes no sense to wonder whether he was smiling or not. And this is not because, although he definitely either was or was not smiling, we could never find out which it was because we did not see it and there is no one we could ask. It is because characters in dreams have no life but the life that shows in the dream. In this they are like characters in plays or novels. It is not to be assumed that Peter and Charlotte either did or did not meet again after the end of the story, and it makes no sense to puzzle over which way it was.

It is not that dream persons have at most only front sides. A dream person is not faceless when his back is turned, and if he says he turned away to hide a smile we would not disbelieve him, or not on general principles. We *might* have a dream in which the unseen parts of people were non-existent; but that would be a novel sort of dream and would be reported with special fascination. In it this curious feature would show, for example, when people leaned against a wall, or had to turn their backs to you to retrieve a wallet from a hip pocket. Normally, however, a dream, like a story, is about people like us, with all their parts there all the time. But while having a face is part of what it is to be a person, having some particular expression is not, and therefore if we imagine a person with his back turned, he *of course* has a face, but it is absurd to wonder whether he is smiling.

One could say: dreams do not have depth. There is no more to be discovered about a dream than we did experience. But this is not the same as saying that there is not more to be remembered about a dream than we do remember.

Or we could put it this way: it is a mistake to try to remember a facial expression on the assumption that everyone at all times wears some expression or other, but not a mistake to try to remember it if one has some specific reason for thinking that it figured in the dream, for example, remembering having been charmed or vexed by it, but not right off remembering what it was.

I said above that one *can* in a dream have the experience of being charmed by a face without the experience of the face itself, and this might seem to imply that remembering being charmed is not a reason for trying to remember the expression. But we would not want to lay it down *a priori* either that the expression that pleased you was not actually there, or that if you could not immediately remember it, it was not there. These are surely empirical questions, albeit peculiarly murky ones. One tries to remember, and then perhaps *remembers* that it was 'not there.' If one tries and fails to remember an expression, one can not conclude that there is nothing to be remembered, but only that one can not remember. What would show that there was nothing to be remembered would be one's remembering that there was nothing.

v *Is there any justification, practical or otherwise, for raising questions about the correctness of dream recollections?*

In the ordinary course of events, other people virtually never and we ourselves hardly ever raise questions about the correctness of our dream recollections, and I would not want the strenuousness of my defense of the possibility of doing so to suggest in any way that we do or ought to engage extensively in such enquiries. We find accounts of dreams entertaining, intriguing, or terrifying, and we envy or pity people for the dreams they have; but generally we would no more enquire whether they do have them than we would enquire whether a person who says he would have liked to stay longer *would* have liked to stay longer.

Similarly, if someone says he dreamt last night of being chased by a tiger, it is somehow off colour to ask whether he could feel the wind on his face as he ran, whether he perspired much, how he decided it was a tiger. These questions are not asked in a normal conversation, and this shows us something of *some* interest about the telling of dreams, namely, that as with the reporting of other incidents, we do not ask questions about facts which do not affect the *drama* of the tale.

There are, indeed, circumstances in which we conclude that a person did not have the dream he purports to have had, but is only saying so to make conversation, to flatter, or to display his attitude toward something or someone – and then we play it differently with him. But (a) even there, we would not generally *contest* whether he had the dream, and (b) this is a different case from the one in which we take it that a person is not pretending, is telling the dream as he remembers it, but in which there might be a question as to whether he misremembers it.

We would not generally have any *grounds*, in the latter kind of case, for suspecting error, for suspecting that the conversation with Ryle that I remember might in fact have been a conversation with Plato, or with nobody in particular. We do not yet know that in certain phases of the moon Ryle does not appear in dreams, while Plato does, or that California is rather too far from England to permit a resident of California to dream of a resident of England. Nor could we, for example, suspect that it was not Ryle because I had him saying things that Ryle would never say; or enquire particularly whether it was Ryle because of the importance, for instance, for the interpretation of *The Concept of Mind*, of his having said what I had him saying.

In these and perhaps other ways it can be seen to be generally an idle suspicion that I might be wrong. But only generally so: we can imagine people (psychiatrists?) with a more extensive knowledge of the ins and outs of dreams and of the consequences of the fact that I dreamt this exactly, rather than that, having very good reasons for enquiring more particularly as to what one dreamt.

It is sometimes said that what *matters* about dreams is only that it should now seem to the dreamer that he so dreamt. But there is no reason why it should not also matter whether he did so dream – why different inferences could not be drawn from the fact that it falsely seems to a person that he had a certain dream than can be drawn from the fact that he did have it. Might there not be a class of psychological abnormality whose symptoms were tendencies to misremember dreams in certain ways?

VI *Have these questions been about dreams, or about the concept of dreaming?*

In many ways it may look as if we have been investigating the *phenomenon* of dreaming, rather than the concept. We have noted various things about dreams in the course of the discussion, such as that it is possible in dreams to take a person to be Peter who looks like Mary, that people may say things in dreams they would never say in real life, that it is possible for dream events to occur neither indoors nor outdoors, possible to have a conversation with a person whose hair is neither long nor short, brown nor blonde. But although we have noted these things about dreams, our problem is not whether these observations are correct: we take them to be obvious. Our interest is in such questions about the *concept* of dreams as whether it is a mistake to assume that a person in a dream had some facial expression or other at every moment in which he figured in the dream.

You might think that anyway we answer these questions by reflecting on how dreams *are*: by recalling that one can have dream conversations with faceless or otherwise indeterminate persons; and therefore there is no distinction between the concept and the phenomenology.

But it will be found, I think, that the key conclusions arrived at in this paper rest on no other difference between dreams and waking experience than that the former are dreams – that is, are not real. It may be that a dream Ryle would say things that Ryle would never say; but even if the Ryle of my dreams is very lifelike, his having made any particular remark is no evidence as to how *The Concept of Mind* should be interpreted, *just because it is a dream*. Similarly, it may be that we can have conversations in dreams with faceless or figureless persons, but in the most lifelike dream it makes no sense to wonder whether a person whose back is turned is smiling, not because we find it to be the case in dreams that persons whose backs are turned have no facial expression (if persons in the same dream so situated as to be able to see these people's faces so assured us, it would make no sense to *believe* them), but just because they are dream figures, and as such, without being regarded as in the least weird in the way dream figures *can* be weird, are not to be thought of as having any *depth*, any characteristics that do not show. (If Ryle behaves strangely one day, one may wonder whether it is because he has a splitting headache. But if the man in my dreams behaves just the same way, one may not wonder this. Except perhaps *in the dream*.)

That it makes no sense to ask such questions is not something we discover about dreams, but something we bring to dreams. We can quite imagine a race who thought they lived in one world by day and another by night, and that although the night world had some special features, such as discontinuity and want of coherence, it was in general of the same logical kind as the day world. The couple of whose romance they dreamt last night either got married after the dream ended, or they did not; the man glimpsed on a far hilltop was either admiring the view, or he was not. We can imagine them finding it deeply regrettable how many unanswered questions their night life left them, and that some of them would spend endless hours trying to fit together the pieces, and to devise ways of further exploring the night world. We would laugh at them for believing in such a world, but if it came down to a hard argument there would be nothing about dreams themselves to which we could draw their attention and which would show or even suggest that they were wrong. If we ever convinced them it would not be by parading facts about dreams, but by invoking such principles as Occam's razor; but of course it would be unlikely that such people would be very partial to that principle.

Telling

This essay takes as its starting point the following remark of Wittgenstein's:

'But when I imagine something, something certainly happens!' Well, something happens – and then I make a noise. What for? Presumably in order to tell what happens. – But how is *telling* done? When are we said to tell anything? – What is the language-game of telling?

   I should like to say: you regard it too much as a matter of course that one can tell anything to anyone. (PI, §363.)

   I start from here, but I will not make it my business to ascertain or guess how Wittgenstein did or might have answered the questions this passage contains. Rather, I will try to answer these and some related questions myself. The answers I arrive at might serve as a kind of guide to the interpretation of Wittgenstein, but might equally serve as a criticism of him, if my conclusions seemed well-founded, and it was shown that they differed from his. I do not care which way it works out, nor do I have any strong views as to how far I am in agreement with Wittgenstein on these matters.

I ON THE MEANING AND USE OF 'TELL'

The sense of 'tell' that we will be discussing is not the sense it has in questions like 'How does (can, should, might) one tell whether ...?' where 'tell' means something like 'ascertain,' 'distinguish,' or 'decide'; nor will we be discussing all of the cases of telling that remain when we have set aside this 'ascertain' sense.

The following are examples of most of these remaining cases: we talk of telling (unfolding, spinning) stories and jokes, telling lies and the truth, telling (ticking) people off, telling (informing) on people, telling (relaying to) people that such and such, telling (ordering) people to do things, telling (ascertaining *cum* spinning) fortunes, tell (reciting) one's beads, and telling (revealing) secrets. There is also an interesting emphatic use of 'tell,' as in 'I tell you, that is not the man,' or 'I tell you, I didn't do it,' where the suggestion is that it is not a matter to be argued over – the speaker knows and is saying, and that should settle the matter.

While I will have some things to say about some of these uses, the kind of case upon which I wish to concentrate is the use of 'tell' in referring to certain sorts of expression of feeling or attitude, or to declarations or avowals that explicitly reveal something about the speaker. When people say 'I love you,' 'I have been thinking ...,' 'I think you are a fool,' 'I hope she will come,' 'I intend to go soon,' 'I did it because ...,' they may afterwards describe themselves or be described as having *told* (someone) that they love her, they have been thinking ...

It will be noticed that, while everything that I am proposing to discuss can be described (and as I shall later suggest, is best described) using the word 'tell,' the word 'tell' is not itself used in the telling. These are cases of telling, but not cases of the use of the word 'tell.' We are, one might say, going to discuss the phenomenon or activity or performance of telling, rather than, for example, the meaning or the use of the word 'tell.'

It is not, however, a peculiarity of the phenomena we will be discussing, setting them apart from other cases, that the word 'tell' is not normally used in the telling: the same is true of telling a lie or a story or a secret or one's beads, and of telling people what to do and how to do things. If a justification is needed for confining the discussion in the way proposed (or is it, for describing what we are going to discuss as a discussion of *telling*?), it is that the word 'tell' is particularly difficult to replace with another word or expression when it is used to refer to these personal revelations or avowals. The main pressure for the retention of the word 'tell' in many of its other uses is not one of sense, but of elegance: because 'tell' takes a person word as a direct object ('tell us,' 'tell him,' 'tell the court'), these constructions are less cumbersome than 'describe to us,' 'explain to us,' 'report to us'; but in these cases generally there will be a locution available employing one or other of such words as 'state,' 'describe,' 'report,' 'order,' or 'say' which, however inelegant, has the same sense as the 'tell' construction. This does not seem to be so, however, when a personal revelation or a personal expression is described using the word 'tell.' A girl could not, instead of saying 'Charlie told me last night that he loved me,' say that he so stated or so reported, except for some such special purpose as to express her contempt for his declaration, or perhaps to capture the excessively formal way in which he had delivered himself of these tidings – in either of which cases there would be a difference of sense.

Similarly, 'He said that he intended ...' is overtly non-committal as to whether he did intend, while 'He told me he intended ...' expresses a measure of confidence between teller and person told. One would express oneself this way when either one had no doubt as to whether he did intend, or when one was surprised that someone else should express that doubt, and was reluctant to doubt that he intended.

It might be suggested that because 'tell' is unusual amongst speech-act words in taking person words as direct objects (it is not unique: 'inform,' 'advise,' and 'order' are like this too), we use it to mark a kind of solidarity between persons that is to be expected in situations of intimacy or in situations in which, for whatever reason, we must rely on the say-so of the speaker. Personal revelations and avowals are, of course, a prime case of this necessity of reliance on the speaker's say-so, but there are other cases.

A witness is asked to tell the court what he saw, or what happened; and this is a situation in which although he may speak falsely, the court has no other resource on which to base its decision than the testimony of this and other witnesses.

Again, the prime minister, or other policy-maker, may tell us what his policy is (and not what he believes his policy to be). He may tell us falsely, but since his say-so makes something policy, if he does other than he told us, he is going back on his word. We are entitled to rely on his say-so, and when such reliance proves to have been foolish, we charge him, not with error, but with deceit, or with inconstancy.

It may appear different with a political pundit, telling us what he believes the prime minister will do: he may turn out to be wrong. But the telling in his case is as to what he believes, and about this he can scarcely be in error, although he may, like the prime minister, be deceitful, or changeable.

The emphatic 'I tell you ...' is a sort of inversion of the trust situation:

whereas in other cases the hearer, in using the word 'tell,' is signalling his trust of or his entitlement to trust the teller, in this case the speaker is demanding to be trusted, expressing outrage at any suggestion of distrust.

What I have called the 'confidential' feature of the way we use the word 'tell' is in some ways unusual, and therefore hard to specify clearly. It has to do, not so much with anything that is true of the person who is referred to as telling, but with the perspective of the person using the word 'tell.' He is saying that he saw the situation as one of confidence, one in which he was entitled to have no doubt. It is not the case that we may not disbelieve what someone has said when we see it as a case of telling, or as some philosophers would put it, that the teller's say-so is a criterion of the truth of what he says. Rather (a) the person describing the saying of something as telling regards himself as particularly entitled to expect truthfulness; (b) the falsity of what one is told is regarded as being necessarily a case of deceit, rather than of error; (c) the truth of the matter in question is regarded as something to be settled, if at all, by the teller's say-so. If the prime minister misleads us as to what his policy is, he is, nevertheless, the only person in a position to say what it is, and we are entitled to hold him to his words, whether deceitful or not; if the witness misleads the court as to what he saw, nevertheless only he can give his testimony; and generally if a deception is uncovered, it is a more bitter complaint to say 'But you told us ...' than to say, for example, 'But you said ...'

Because of this very general tendency for 'tell' to be (in the sense indicated) a 'confidential' word, it is perhaps no accident that it is so naturally used, and indeed so awkward to replace with other speech-act expressions, in referring to personal declarations, avowals, and expressions of feeling and attitude. In these cases *par excellence*, it is for the speaker to say. In other cases such special circumstances as a person's being a witness in a courtroom or being the person who makes policy put him in a 'confidential' position; whereas in the cases we are to discuss he is in that position regardless of the circumstances. Only I can say what I have been thinking, why I did it, what I dreamt, what I hope, what I intend. In so saying, I will not necessarily speak truthfully, and you may disbelieve me. But you may not contradict me: you may not talk as if you knew better than I, what I intend for example.

When you challenge what I say, the aim of your challenge will be, not to demonstrate to me that I have made a *mistake*, but to suggest that I have been disingenuous, and hence induce me either to make some alternative profession, or to commit myself more particularly, by further professions, to what I have already said.

You may, for example, say 'If that is your intention, why did you tell Mary that you would not do that?' – and then I may confess that I lied to Mary, a further profession of mine that permits me to stand by my original declaration of intention. But again I may break down and admit that I deceived you, in which case again you will have shifted me from one avowal to another. That is the characteristic pattern of challenges to these tellings.

This pattern is to be compared with something one may not do: one may not say 'You only think that is your intention. I will prove to you that in fact you intend thus and so.' Another person may, it is true, know better than I do what I will in fact do: he may be much clearer than I am as to what my weaknesses and propensities are, and hence be able to make a sound prediction as to my future behaviour. But such predictions will not contradict any statement of mine as to my intentions, because in declaring my intentions I am not making enlightened guesses as to what I will do. Predicting is a different business from declaring an intention, even if in many ways it is related.

If you contest my *prediction*, we may jointly assess the evidence; but if you disbelieve my statement of intention you go to work, not on the question what is true, but on *me*. You try to bring it about that I retract or reaffirm my intention. In the event that I do reaffirm it, it is not the case that you now *know* what I intend, but rather that you have induced me to commit myself more particularly, making it more awkward for me later on to disavow that intention.

In this way, scepticism about a person's intentions and discussion pursuant thereto trade always in further declarations, professions, confessions of the intender. It is throughout treated as being for him to say. He is treated as being in a certain way privileged. 'Only he can know what his intention is' is an expression of this recognition of privilege.

Whether or not the foregoing examples demonstrate it, I am going to assume that at least in the kinds of tellings we will be discussing, we rightly accord the teller an ultimate privilege: not the right to be believed whatever he says, but a right that could be expressed in the proposition 'Whatever he may in fact say, at any rate only he knows.'

If now we ask, why is this privilege accorded, an answer to which we may be attracted is 'Because what a person tells us is a report of something that is inaccessible to us, namely his inner feelings, thoughts and processes. Since only he has access to these, while he may deceive us as to what they are, still only he can set the record straight.' There is, however, another answer that has appealed to many people, namely, that the privilege is based on the speaker's access to *more* of a kind of information that is in principle accessible to us all: information about his behaviour.

Much of this essay will be concerned to show that neither of these answers is correct. It will then seem pecularily difficult to say what the basis of the privilege could possibly be; and it will be this apparently baffling question that I will finally be attempting to answer.

II ON THE 'LANGUAGE-GAME' OF TELLING

I take the question what is the language-game of telling to mean: what sorts of response to a speech-act that we would describe as a telling are appropriate and inappropriate? In what characteristic ways does a conversation in which someone tells us something proceed? But if this is not what Wittgenstein meant by his question, it is in any case the question I now wish to ask.

I will contrast the language-games in which tellings of the kinds we are discussing figure, with the language-games of reporting one's own states or processes, whether physical or mental.

The moves in the latter kind of game, I wish to say, are centred on the biographical events themselves: on their exact description, the course they run, their duration, their time and frequency of occurrence, the conditions under which they occur, their side- or after-effects, and our reactions to them of pleasure, annoyance, discomfort.

If I say I have been having a funny feeling in my lower abdomen, or that my mind will not stop turning over at night and I can't sleep, you may want to know more about the funny feeling, whether it is more like this or like that, whether it is painful or merely disconcerting, whether I have it all day or perhaps only after eating; or you may want to know whether what runs through my mind is verbal or pictorial, coherent or incoherent, unpleasant or enjoyable, and whether it is connected with my daytime concerns and anxieties or whether it is merely random fancies.

When I tell you what I have been thinking, what I intend or what I hope, by contrast, the pursuant conversation centres, not on what has been happening to me but, one might say, on the upshot of what I have said: not on the thinking by me of something, but on *what* I have thought, not on my intending but on what I intend.

If I say 'I intend to go to the meeting,' you may say 'Oh in that case I'll go too,' or 'I wouldn't if I were you. I think it will be dull,' or 'How can you go? You have a lecture at that time.' You start from *what* I have told you (that I am going), and go on from there.

You do not venture that it is annoying when these intentions happen; or suggest that perhaps if I smoked less I might not find myself intending so much; or enquire whether having intended something, there is any way I can prevent myself doing it; or whether intending in my case is just like it is in yours, and if not what makes me so sure it is intending. You do not in short treat my declaration of intention as a report of an interesting mental event. You take an interest only in what is intended. You start from there: you express your approval or disapproval, or you invite me to reconsider, or you probe the problems and prospects that arise out of the intention.

If I tell a student I think his paper is excellent, he is pleased and perhaps he thanks me; but he is not thanking me for the performance of some activity of so thinking, but for the compliment. It is not as if the activity of thinking his paper excellent, though difficult and perhaps unpleasant for me, was generously performed, and would do him some good. (We might thank people for thinking nice things of us if we believed that, like silent prayers, they stood a chance of benefitting us.)

If I tell my wife that with any encouragement from her I would quit my job tomorrow, I am certainly revealing to her something about myself: the extremity of my dissatisfaction with the job and my anxiety to do nothing not having her support. But I am not showing her what has been going on in me. She may be impressed that I should be able to make such an extreme declaration; but it will be just that, that she is impressed by. In her reaction she will *go on from there*: she may suggest that if I can just hang on a while, things will get better, or she may argue that there are just no available alternatives, or she may say that although she is terrified by the prospect, she will support me.

Such facts as the foregoing about the role of these tellings in human interaction suggest that whether or not they appear to do so, such tellings do not describe, report, or allude to episodes, processes, activities, or states in the life of the teller. One could, however, agree that the role of these tellings in human intercourse centres on the upshot of them – on the compliment that is paid or the stand that is taken or the expectation that is created – but still hold that they are nevertheless in part biographical remarks, and that therefore even if it is inappropriate for *participants* in language-games of telling to be interested in them as biography, it is not a mistake for other people, philosophers or psychologists perhaps, to take an interest in their biographical aspects. In Part III I will argue in detail that no biographical investigation of these tellings is warranted; here I would like to offer a general argument to the same effect.

Only in some cases of telling does there seem to be an *explicit* biographical statement made. 'I have often thought ...,' 'I have been expecting her all day,' or 'I have been intending to write for a long time' seem to make such a statement; but not 'I think that is silly,' 'I wouldn't go if you paid me,' 'I intend to write tomorrow,' or 'I hope she will come.' Consider a case of the latter kind. If I tell you what I think of you, then while I am of course revealing something about myself, namely what I think of you, there seems no way of saying what I am revealing that shows it to be some particular or set of particulars from my biography. I am not saying that the mental process of my thinking you a cunning rogue has occurred at least once, or that it tends to occur whenever I see you or whenever your name is mentioned. It can be true that I think you a cunning rogue if I have only now met you or only now come to that conclusion about you, and if my so telling you is the very first expression of that opinion. And even if there has in fact been a series of similar thoughts or statements about you, the items in this series could not always be based on earlier items; and if any one of them can exist without the backing of the others, then so can every one. Each can be an original expression of my reaction.

Let us look then at tellings that seem to make an *explicit* biographical statement. If I say 'I was just thinking ...' or 'I was thinking only this morning ...,' something has happened, at a (fairly) definite time. Are we not entitled to enquire at least in these cases as to the exact nature of the event or process referred to?

What might we find upon investigation? The first point I wish to make is that the plain facts in every particular case will always be open to various interpretations. If my wife complains one day about the condition of our car and I say 'Yes, it is in terrible shape, and I was thinking only this morning that we should get a newer one,' what may have happened this morning is that I felt particularly vexed by the car's rust and rattles, tried to remember how much money there was in the bank, reflected a bit on foreseeable expenditures of money, and said emphatically to myself 'We must get a newer car.'

These events are consistent enough with my later saying I was thinking we should get a newer car; but they are also consistent with my later saying that I would of course like a newer car, and that *only this morning I was trying to talk myself into getting one*, but that really I do not think it would be wise at this juncture. Hence I could not, just by remembering what happened this morning, make this afternoon's representation about those happenings.

This is not to say that I would be entitled this afternoon to put various interpretations on what happened this morning, depending on how I was now disposed. Only one account of what I thought this morning can honestly be given; but if the plain facts are consistent with various accounts, what is to show which is the honest account of what I thought?

The answer that I wish to suggest is: *what I am now prepared to say*. If there is a doubt as to whether I favour the purchase of a newer car, the fact that I am prepared to go on record as favouring it, to take this step that will seriously restrict my room for manœuvre, shows if anything does that I favour it (this afternoon); and if my attitude has not changed since this morning, then, in saying to myself then, 'We must get a newer car,' I was not engaging in a reverie or trying to talk myself into it, but was thinking that we must.

Here I do not mean that this afternoon I reason in some such way as the following: 'I must now favour a newer car, because I am altogether ready to go on record as favouring it; my attitude has not changed since this morning; therefore what I said to myself this morning must have been a case of thinking that we should get a newer car.' If I am in doubt as to what I thought this morning, if there was any need to reason about it, that would be evidence enough that what happened this morning, whatever it was, was not a case of my thinking we should change cars. If I did think this morning what I now say I thought, I knew at the time that I did; but neither then nor later do I know by inspecting what emerged, but rather by being the person who generated what emerged. I am able to say now what I thought because I am the same person.

The following might be suggested as an objection to this thesis: if it is what we are prepared to say that shows what we thought, how is it that we can know what we were thinking in the case in which we have since undergone a change of attitude, and are no longer prepared to go on record in the same way as we would have done earlier? Suppose, for example, that since this morning I had remembered a mortgage payment that would soon have to be made, and hence no longer felt that we could manage another large expenditure. I would hence not now be able to say that I favoured a newer car, and yet I could say that I favoured it this morning.

Two replies to this objection could equally well be made, I suggest. The first is to say that in that event there would ordinarily be no purpose served by excursions into my history, and I would do better just to declare my present attitude. The topic in our example after all is cars, not psychological events. There is not normally any need to settle whether events in my history are to be seen as daydreaming, self-persuasion, or the expression of a firm attitude. But (secondly) if in some circumstance I found it useful to say 'Yes I thought that only this morning, but at the time I had forgotten about the mortgage payment, and now ...,' I would be going on record as conditionally favouring a newer car. Having so declared myself, I would have little room for manœuvre if it turned out that I was mistaken about the mortgage payment. This qualified commitment would be sufficient to mark this morning's events as what I said they were.

Our question has been whether, although for participants in language-games of telling only what we have called the 'upshot' is important, there may nevertheless be a question that philosophers or psychologists could answer as to what events or processes in the life of the teller are being revealed when he tells us something. The foregoing discussion has brought out the following points in answer to that question:

1 In saying what one has thought, one is not even hinting at what happened. A very large number of different things might have happened (although no doubt one could not properly say 'I have been thinking ...' if nothing had happened that fell within a certain very large range of events).

2 The discovery of what in particular happened will not itself show whether the teller's account of what he was thinking is a fair one.

3 It is not on the basis of the teller's access to what happened that he is able to say what he thought.

4 Nevertheless, he is not at liberty to put any construction he wishes on what happened. What happened this morning either was or was not a case of his thinking such-and-such, and whichever it was it is disingenous of him to describe it otherwise.

5 If the teller is in any doubt as to whether, for example, he was day-dreaming, trying to persuade himself of something, or thinking something, that is evidence enough that his attitude at the time was at any rate too

indeterminate to warrant his now telling us anything so definite as that he was thinking thus and so.

6 What shows what he thought is what, realizing what stand he is taking, what expectations he is authorizing, from what subsequent moves he is foreclosing himself, a person is now able to mark himself as thinking.

7 A person does not *conclude* from what he is willing to say this afternoon what he should make of this morning's events; but being the same person who thought something this morning, he is able to say this afternoon what he thought. It is not by examining and interpreting what happened, but by doing more of the same, that one says what one thought.

The following analogy may illuminate this last point. If I have made a somewhat complicated point in a discussion, and perhaps not made it very successfully, and am afterwards telling someone what point I was trying to make, the most useful procedure will be, rather than either repeating my actual words or offering some direct translation of them into other words, just to start afresh, to reattempt the point. Clearly that is possible, it is what we usually do. If we ask how it is possible, essentially the answer will be that, there being something that I understand and my having certain abilities in explaining things, I re-employ those abilities to re-explain the point I had made. It is not the case that there is something definite there that I experience and work up into an explanation by applying to it certain rules or techniques; the explanation flows not from any such something but from *me*. How do we know it is the same point that is made on each occasion? Not because both explanations start from the same experience, but because they are given by the same person as the same point. That person may, with or without realizing it, shift his ground; but that will be shown, not by reference to something from which he began, but by comparison of the two explanations he has generated.

To have thought something this morning is to be a person who could have gone on record as so thinking. From this nothing follows as to what this morning's events must have been in order to be a case of thinking such-and-such. About that, there is nothing to say.

III HOW TELLING IS DONE

In Part II it was argued that in the language-games in which tellings figure, what is important is not the biographical events or states (if any) that are

told of, but the 'upshot' of the telling, the reactions and subsequent developments in the language-game that become appropriate and inappropriate,
given a certain telling. Further it was argued that this is true not only for
participants in the language-game, but for anyone: there is nothing that
shows whether anything that happened is properly described in the way
the teller does, except the fact that, knowing what he is getting into by his
so delivering himself, the teller is able to do so.

The present section will further bear out the latter of the above claims,
by arguing in some detail that we do not in fact arrive at what we tell
people about ourselves by close scrutiny of ourselves, for example by ascertaining exactly what happened and then determining whether what
happened conforms to some model, or satisfies some criteria, of thinking,
intending, hoping, or expecting.

There will be two broad types of case here, those in which one is telling
of events or sequences of events that did occur, and those in which nothing
is represented as having occurred, but in telling we are, for example,
explaining ourselves or expressing ourselves. It will not always be very
obvious into which of these classes any given case falls; but that question
may for the present be avoided if we take for examination tellings that are
clearly members of either one of the two classes of case.

1 Amongst tellings that reveal something about oneself, nothing would
appear to be a clearer case of telling *what happened* than that of telling someone one's dream. I will therefore attempt to show two things about dream-
tellings: (i) that although the telling here is a description of what happened,
it is not done by reading off the description from anything, that is to say
we are not, in telling the dream, like an on-the-spot reporter who constructs a running verbal account of what he sees happening; and (ii) that
in spite of the dream being something that happened, what I said in general
about the language-games of personal tellings is true also of dreams.

It may be natural to suppose that if something like a dream occurred, we
will tell people what we dreamt by (in some way) rerunning the dream and
giving a commentary on what we then observe. But right away we can see
that there will be many difficulties in this supposition. How will we know
that the rerun resembles the original dream? Not, surely, by *comparing* the
two, because the dream is gone by, so that the only available thing with
which to make comparisons might be a further rerun, which, however,
would carry no better guarantee of authenticity than the first.

If we avoid this difficulty either by supposing that we 'just know' that
the rerun is authentic, or by saying that we do not ask whether it is, but

simply start from there, there will still be questions as to whether the replay runs through at the same pace or in the same sequence or with the same richness of detail that we end up saying it did, or if not, on what basis we make modifications involving the pace, sequence, detail? If we had only the rerun to work with, we could have no reason to do other than describe it exactly as it comes; but if we know independently of the rerun what we should say, then we no longer have any reason to suppose that the rerun is our source of information.

We are likely, however, to feel that there must somehow be an answer to such difficulties, because (many people would say) we are in fact conscious of reliving the dream as we tell it. The question, however, is not whether this is (always) the case, but whether even if it is, what we say is *derived* from what we are experiencing – or whether perhaps we say how it was, and because we are so saying a likeness of how it was (sometimes) appears. What would show whether we *use* the imagery in arriving at what to say would perhaps be (i) whether without it we are lost; (ii) whether we are often surprised by it as it develops, as you might expect if we did not know independently of it what had happened; or (iii) whether we *study* it as it emerges, to find out exactly what to say.

But (i) it is not clear that we are lost without it; there are sometimes gaps in the imagery without our being uncertain what happened at those junctures; we sometimes get ahead of the imagery; we are sometimes sure that more happened than is included in the imagery, or that it happened in a different sequence or at a different pace. (ii) We do not usually tell people our dreams with the kind of curiosity, surprise, and disappointment that we might if they were happening now and we were wondering as they unfolded how they would turn out. If we say 'and to my astonishment he turned into a crocodile and slithered away through the long grass,' we are not now astonished, but are reporting an astonishment (for which, incidentally, it would be difficult to suggest a mode of appearance in the rerun). (iii) We are sometimes at a loss how to describe something in a dream, but not because the imagery is not clear enough or has not been closely enough examined. If I were to have a closer look in order to answer some question of detail, say as to whether a smile was supercilious, I might generate the picture of a supercilious smile just because my question is whether the smile is like that. If I do remember what kind of a smile it was I will perhaps concurrently picture it, but it cannot be from the picture that I derive the recollection.

When we are uncertain what happened at some juncture in a dream, or what someone was wearing, or whether a person in a dream looked like the person that in the dream we took him to be, we may do various things to nudge our memory, but if they work what has happened is just what we remember. The blockage is removed and we are able to proceed. We say 'I remember now. Isn't that extraordinary! Although I certainly took him to be Uncle Milt, he had the appearance and manner of Louis St Laurent, and that didn't strike me at the time as being at all odd.' Perhaps as one says this, the dream will come back very vividly; but it is neither the vividness nor anything else about the experience we thus have that shows that one is now remembering, rather than, for example, imagining. One's confidence that that is how it was is independent of the specific properties of the experience one has when remembering.

If we do not somehow derive what we say in telling a dream from some experience, how *is* telling done? My suggestion is that this question wrongly suggests that there are some techniques or procedures that we follow, whereas in fact we are simply the kind of beings that, when we have learned a language and had a dream, then if someone asks us about it or something reminds us of it, we can without more ado talk about it. We do have certain abilities and skills that are employed in the telling: the ability to remember, and skill in making what we say interesting, suspenseful, or intelligible to a certain audience; but normally these abilities simply function, without any conscious machination.

When I said 'without more ado,' I did not mean that remembering dreams (or remembering anything else) is effortless, and certainly not that we will never fail to remember them, or never misremember them. We do become confused and bogged down, struggle to remember, and even suddenly remember that we have been getting it wrong. I only meant that there is no science to remembering; there is nothing we do that is artfully contrived to make it come, or to get it right. We perhaps screw up our face a bit, clap our forehead, or gaze intently at the sky; and then things come clear, or they do not. But these things are not a *method* of remembering: we do not teach people this way of doing it, or suggest that a person cannot be trying very hard to remember, because he has not even clapped his forehead, and that is one of the very first things to be done.

Is the telling of dreams, however, not a counter-example to the claim that in personal tellings it is not what has happened, but rather the upshot, the role that the telling plays in some language-game, that is important? Perhaps it is: we certainly often describe dream events in considerable detail, and it is appropriate to apply for further information about them in a way in which it is not appropriate to ask, for example, just what it was like to think this morning that we need a newer car.

There is, however, a sense in which it is still the 'upshot' that is important. In telling you my dream I am sharing with you the terror or the enchantment or the drama of the experience. That defines the nature of the language-game: only such questions about what happened as help the other person to savour the horror or the pleasure of the dream may be asked. If I tell you I have been chased in a dream by a most terrifying mouse, you may ask whether it was one of those ghastly green ones, whether it growled and snorted, and whether I had that awful sensation one sometimes has in a dream of impotence. But you should not ask whether I was wearing my wrist watch, whether I perspired much, whether the mouse had a tail, or if I am quite sure it was a mouse and not a hamster. Not that I might not be able to answer such questions: indeed, I might have a dream in which, dream fashion, it was just the fact that I was wearing my wrist watch that made the mouse so terrifying. But except in special circumstances you are not playing my game if you ask any of the latter questions, because it will not be clear how the answers to them will help you to capture the *interest* of the dream. We tell dreams as we tell stories: to beguile people; and the two language-games are similar. One is not entitled to disbelieve a dream or criticize a story because it is incredible; one may suggest wrinkles or features for a story or a dream if they would contribute to the enchantment, but it is in both cases for the teller to accept or reject these. It is his story, and it is his dream. The dream, however, did happen, while the story normally did not; and among the differences this makes are, first, that when the dream-telling breaks down we have forgotten, whereas when the story-telling breaks down our ingenuity has failed; second, that one may mention features of a dream although they make no contribution to its beguiling power, and just because that is the way it was; and, third, that there may be other kinds of interest attaching to dreams, such as clinical interest, and hence there may be a switch from the language-game of beguiling to that of diagnosing, in which a different set of reactions and questions becomes appropriate.

2 We have so far been considering how telling is done in cases where what is told of is something that actually happened. The question will work out somewhat differently in cases where what we tell of is *not* something that happened: cases, for example, of telling someone why one did some-

thing, what one meant, that one would have liked to stay longer, or what one was hoping, intending, expecting. I shall assume, without arguing the matter, that although most of such tellings are about a time past, they are nonetheless not about actions, events, or processes in the past. There may, at the time referred to, have been some expressions (verbal or otherwise) of the intentions or expectations, but what we mean when we say that we expected or intended is not that those expressions occurred. If, having looked at my watch and realized that we had to leave the party, I said to my wife 'I would like to stay longer, but we must go,' and she sadly agreed, my later remark to our hostess that we would have liked to stay longer is not a way of saying to her that I said some such thing to my wife as I did say and that she agreed in the manner she did. I could say we would have liked to stay longer if nothing of that kind had occurred; and I could say it untruthfully even if that exchange between myself and my wife did occur.

For this type of case one might be inclined to offer a more sophisticated account of how telling is done. Or perhaps it is not that it is more sophisticated so much, as that one may be driven to this kind of account if one insists that there must be a *procedure* for arriving at what we say in cases where, since the telling refers to no events, the procedure cannot be that of describing them.

A knowledge of various facts about a person will very often enable us to tell (in the 'ascertain' sense) what a person meant, hoped, expected, intended. For instance, if I said something that could be taken in various ways, and the person to whom I said it took what I said to be a sarcastic remark about him, then anyone who knew such things about me as that I like and respect this person, and that I have a certain kind of sense of humour, of which my remark can readily be seen as an example, would have reason enough to say 'You don't understand: he meant it as a joke.'

In this and similar ways, other people can often make judgments as to my attitudes, motives, intentions. It might therefore seem plausible to suggest that I myself, when I tell you what I intend or why I did it, arrive at what I tell you by reflecting on facts about my tendencies, tastes, interests, actions, and circumstances.

Perhaps the most serious of many objections to this supposition is that it requires us to regard a person as *ascertaining* how he meant it, what he intends, why he did it, and if one needs to ascertain, one can be more or less certain of *what* is ascertained. We ought therefore on this supposition often to find ourselves saying such things as 'I suspect that I meant it as a joke,'

'It is virtually certain that I would have liked to stay longer,' 'All the evidence suggests that I intend to go.' Now each of these utterances when placed in the context of an actual transaction between persons, for example, the transaction of placating someone who has misunderstood my witticism, or that of making a cordial departing remark to one's host, will look extremely odd. And the reason for this is that these tellings have the character of *assurances*. (One *could* always say '*Believe me*, I would have liked to stay longer,' or '*I assure you*, I meant it as a joke.') And it is ruinous to an assurance if it includes any expression of uncertainty.

There are other difficulties. If there is a method of reviewing circumstances, deciding which of them are relevant and which not, and deciding between any two or more inferences from such evidence which would more or less fit the known facts, the method would be a human invention, and it would be a good one only if it seemed generally to produce results that were *right*. But this would require that it should be possible to know independently of the method what a person intended, expected, and so on. The method could not be *more* reliable than the other way of knowing: if, following the method, someone concluded that I meant a remark as a joke, but I denied it, it would neither follow from this that I was lying, nor could any amount of rechecking of the evidence or of further research show that I was lying. Not that I might not *be* lying, but until, for whatever reason, I admitted as much, there would be as good reason for saying that the method had yielded the wrong conclusion, as the contrary.

You might be able to get me to admit that I was lying by getting me to play the method game with you: by setting out your reasoning and asking me to point out flaws in it or to adduce evidence that you had overlooked. If it happened that when every move I made in this game failed, I admitted that I had been lying, that might seem to show some kind of primacy of the application of the method over what I professed or admitted. But I could agree that all the evidence pointed in a certain direction, and still assure you that it was the wrong direction. As was suggested in Part 1 (pp. 94–5), you would not then *know* that I had not lied, but you would have induced me to make various admissions and further commitments. You might, for example, have foreclosed me from later saying what might otherwise have been open to me to say – that I had only been joking.

We cannot get around this difficulty by saying that each man knows privately what he means or intends, and that this is the independent knowledge that has made the development of a method possible. It is just the absence of any obvious way of 'knowing privately' (because intentions etc. are not actions, events, or processes), that has led us to posit the method; and it was supposed to be a method to be used by oneself as much as by other people.

It is not even clear that it is a method for other people. Someone takes my friend's ambiguous remark to be sarcastic and I immediately feel that that is wrong. I would not trouble to demonstrate this to myself; but how should I convince the hurt party? Well, I cast around and perhaps find evidence that I had not thought of when first it seemed to me that he was wrong. I manage to find a letter that I had not even known of in the first instance, but which does contain convincing evidence; or after racking my brains considerably I remember a remark that serves the purpose.

But whether or not it is a method for other people, one certainly does not in one's own case work out in this way such things as how one intended a remark or whom one was expecting. There is no point at which one is puzzled by the ambiguity of what one said and wonders what one meant, no point at which one is weighing hypotheses or searching for more evidence, no point at which one hypothesis emerges as clearly to be preferred and one opts for it. We seem to be able to say without reflection what we meant. If anyone doubts what we say, we may employ a method to convince *him*, but not to convince ourselves.

Let me note and comment on three considerations that might make it difficult for us to see this. In the first place one may sometimes say that one meant something as a joke, having suddenly regretted a momentary inclination to sarcasm, and out of the sincerity of one's present desire to placate, really and wrongly believe that the remark was not intended sarcastically. In this and other ways it is possible to make mistakes as to how one meant it, and it is reasonable to suggest that the only condition under which it would make sense to talk of errors here would be if there were some method of arriving at the truth. If it is true that there must be *some* method, however, it is not therefore true that the method is the one we have been discussing. Perhaps the method is that of putting it on to oneself in a certain way (the 'honestly now' way), whether one really meant it as a joke.

Secondly, if we are asked about something we have said some time ago in circumstances now forgotten, we will often be unable to say what we meant until we have reminded ourselves of the interests and views we then had, or the context in which the remark in question was made. This may look like at least one kind of case in which we decide from a review of the circumstances what we meant; but I think we misunderstand the mechanics of the review if we so conclude. It is not the case that anyone upon reviewing the things we survey would be driven to the same conclusion. The review does not show us what we meant, but helps us to *remember*. We refresh our minds about the circumstances, and then the answer *comes* (or it does not). The reminders we give ourselves put us in a position to say right off what we meant, the way we would do if the situation in question had only recently occurred.

Thirdly, we sometimes do not know what to say when someone asks us what we meant, and this may suggest that it is false that we know such things right off and without machination. But when we are thus at a loss, it need not be because we do not know what we meant: we may not understand why anyone would ask. What we *said* may seem to us the only or the best way of saying what we meant, and we may not right away see what we could say to anyone who finds what we have already said puzzling. We tackle the problem, not by examining ourselves or our circumstances, but by trying to ascertain where the other person's difficulty lies.

It has appeared that whether in the case of tellings that are or in the case of those that are not about actual biographical events, there is no answer to the question how we do the telling, that is, there is no procedure we follow, no technique we employ; but with or (more usually) without some effort, experiencing some tension, we simply do it. It cannot be claimed that this has been *demonstrated*, but only that some of the main temptations to say otherwise have been dealt with.

Let me conclude this Part with a further reminder of the importance of the role of these tellings (particularly the non-biographical kinds) in some language-game. We are inclined to think that if I, for example, meant a certain remark as a joke, then although I will say so only if someone misunderstands, it is true that I so meant it, whether or not the occasion to say so arises. Hence in writing my autobiography I might properly record that I meant it as a joke, even if I could not also record that anyone had misunderstood, or that no one had been amused. Yet if the remark that I record is patently amusing, one would be left puzzled as to why I should say I meant it as a joke unless I went on to say that Peter was offended or that literal-minded Jane thought I was in earnest. It is not just that what I record is uninteresting, the way it would be uninteresting to record that I was wearing my wrist watch at the time: there is nothing, not even something very boring, to record. The only function of remarks like 'I meant it as a joke' is explanatory, and when there is nothing to be explained, there is no sense in which they are true.

There will, I think, be quite an inclination to say that even if, as I have been maintaining, tellings of the kind we have been discussing have no specifiable basis in the history of the teller, and normally flow from a competent language-user without machination of any kind, still there must be some way of knowing whether what he says is true; and, further, a way of specifying what we believe when we believe it. For we do regard these things that people say as being true or false. We do not regard the mere fact that a person has told us something as showing that what he says is true, nor do we always believe him; but we believe him in some cases and not in others, and we react differently depending on whether we believe. What ways then, if any, are there of knowing whether tellings are true, and what do we then know?

The first point I would like to make about such questions is that it is not always, or even generally the case that we believe or disbelieve tellings. Most often the question of truth simply does not arise, but a person says something, and we go right on from there. If someone later asks whether we believed what we were told, we will be likely to say 'Of course'; but all this will generally mean is that the question of the teller's veracity simply did not arise. One says 'Of course' because *some* answer to the question whether one believed has to be given, and not either because the case is at all like, for example, that in which one is told a fantastic story and yet believes it, or because the teller's veracity is so impeccable that one would not dream of disbelieving him.

Secondly, I suggest that many of the cases in which a doubt is raised can be seen, not as cases in which we attempt to ascertain whether what the speaker said is true or false, but as cases of ascertaining more exactly what move he is making. If I tell my wife that I have been thinking of quitting my job and she says 'You surely don't mean that' and I assure her that I do, she does not then *know* that I have indeed been thinking that. My assurance settles no more as to whether perhaps I was spoofing than did my opening remark. The upshot is rather that she knows in what way to continue the conversation: whether to treat what I said as a prelude to an amusing evening's chit-chat about how we might otherwise live, or as a probe of her reaction and an opener in a deep discussion of an important step. Her question is *like* the question 'What game are we playing?'; and if I answer 'Poker,' all she can do is play it that way. If I later start to snicker and smile at her agony, and chide her for taking me so seriously, it will not be she who has misplayed the game.

Thirdly, further enquiries as to, for instance, what a person has been thinking will sometimes be more searching than they are with such questions as 'Do you mean it?'. In the above example my wife might ask why and how seriously I had been so thinking. But (a) she will only have my assurance as to the truth of my replies to such questions, and (b) we may note a number of interesting things about the kind of replies I would be likely to give. To her question how serious I am, I might, for example, say 'Well I suppose I would not be likely to quit tomorrow, but with some encouragement from you I would certainly start laying plans to quit.' This would surely be a very useful indication of my attitude; but I would not likely say it on the basis of having recently had just that thought, nor is the seriousness of my thought a quality that I am aware of, and for which, not being able to communicate it directly, I devise this indirect indication, this hint. I do not 'see' in the seriousness of my thought just this indication, nor does my wife gather from what I say what indescribable quality it had. I speak, not on the basis of a self-examination, but as a further expression of my attitude, as primordial as the thoughts I have been having, but geared now to my conversation with her.

One would not ask on what basis yesterday I thought some thought about quitting – how I knew to think that thought. I just *thought* it; 'it came welling up,' one is inclined to say; and there is no reason to ask on what basis I said to my wife that I would not quit tomorrow but ... If there is anything that shows me that I should express myself just this way, it is that, knowing the alarm it may cause, or knowing how it may commit me in the event that her encouragement is forthcoming, I am able to say it. I may indeed have had bolder-seeming thoughts prior to this moment, and then you might expect that to be honest I should now express myself boldly; but unless I am a coward and my wife is formidable, my inability now to express myself strongly may show just how serious my thoughts have been, may show, for example, that I have been playing a half-serious private game with myself.

It is important also that the things we say to one another are not just pieces of self-expression, but are at the same time probes of the other person's reaction, offers, threats, appeals. 'With some encouragement from you I would quit' may be an accurate index of my attitude, but it is also an appeal for her reaction and an offer to quit if encouraged. Similarly, if she says 'I will divorce you if you do,' that may be an expression of an attitude, but it is also a thrust in the interplay between us, to some extent committing her to a course of action in the event that I do quit.

How do I know when to make a promise or to appeal for her reaction? How does she know when to make a threat? Well, we do not know. That is to say there is nothing that *requires* just these moves. (She would not, as if noticing something about herself, say 'I *find that* I would divorce you.' These words would be uttered defiantly, not reflectively.) She and I are in the middle of something, and trying to get on with it as well as may be, and what we say is dictated primarily by that fact. We do not of course say just anything that will serve that purpose, but only such things as, knowing their consequences, we are willing to say; still the primary question at any juncture is 'what move may best advance things at this point?'

The question how we know what to *tell* someone is therefore a misbegotten one: it falsely implies that either we do (sometimes) know, or if we do not it is too bad and a dismal fact about us. We neither do nor should know what to say in these contexts. Indeed, if we did, it would seem to imply that the thing to do in situations such as we have been discussing is not to respond intelligently to the other person and develop the interplay fruitfully, but just to say what is the case, to describe one's current state; and that if doing that should serve to advance the discussion, well and good, but if not, that is if there is nothing to describe that serves that purpose, then there can be no interplay. The determination to say nothing that is not a report on one's states would make discussions impossible by ruling out our reactions, since there is nothing to which reactions correspond, or on which they report. 'You had better not!' does not formulate, describe, or depict a reaction, it *is* a reaction.

V THE NOTION OF A LANGUAGE GAME

Our discussion has in various ways brought out the importance, for a philosophical understanding of tellings, of seeing them as moves in some language-game; and I would like to conclude by setting forth as clearly as I can some of the main features of this difficult notion.

1 We have repeatedly seen how the opening moves of a linguistic

exchange determine what *kind* of further development will be appropriate. It is difficult to suggest any useful principles governing the way this determination works, and I can only point to the way we do understand, in various circumstances, what sorts of move will be appropriate and otherwise. I would suggest that part of what it is to understand what someone tells us is to know what sorts of response to it will be appropriate and otherwise. (One may respond inappropriately just because one understands in this sense, however; for instance, when I tell my wife I have been thinking of quitting my job, it may be a very subtle move on her part to treat what I said as a prelude to an amusing session of conjecture as to how we might alternatively live.)

2 The moves we make do not necessarily or even generally convey information, but rather *affect* the other person: amuse, intrigue, comfort, warn, embarrass, threaten him; or appeal to him, for sympathy, encouragement, advice, or just to make some decisive move. There is not always a sense in which such tellings may be true or false, but even when there is, we say them not because they are true, but because they play a useful role in the linguistic exchange that is taking place.

3 At junctures of the language-game that call for self-explanation two elements are involved: one getting it right, and the other putting it in a way that copes with the other person's difficulty. Philosophers often think of the former as being the main thing, and the latter as being a minor, mechanical business; but I have tried to show how the latter in fact is primary. (If my wife, in an attempt to understand how seriously I am thinking of quitting my job, asks whether I would be likely to quit tomorrow, my history need include no such materials for answering her as my having *thought* of quitting tomorrow. If I say 'Yes, that is quite likely,' or 'Perhaps not, but with some encouragement from you I might,' it may be that I see the matter precisely this way for the first time when I reply. Her question has brought it out that this is my attitude. And if we ask whether, had she not asked her question, it would anyway be true that this was my attitude, there just seems no way of answering that question. There is no obvious end to the questions she may ask or the answers I may devise; but what are we to make of the question whether, had she not asked those questions, the answers would anyway have been there unexpressed; or whether, if I answered her questions honestly, then had I been asked simply to spell out my attitude in full, I could have been expected to say at least all those things?)

4 The game can develop only in terms of the moves that have been made. By this I mean two things:

i What anyone says in the language-game *is* the move he is making. We can no more ask whether he has made the move than, if someone puts his piece on a certain square in a game of chess, we can so ask. There *is* a question, that scarcely arises in most games, as to just *what* move has been made (this, I suggested, is what is going on when we ask 'Do you mean it?' or 'Do you mean such-and-such?'), but once that is clearly answered, the game can only proceed from there. If you say something and I do not believe you, the game may end there; but it need not: it may take a new turn, in which I devise a strategy based on my scepticism, perhaps saying just the things you will not want me to say if you are insincere in what you have said.

ii Just as in chess, for example, I may make a powerful move by accident or with some other strategy in mind, so the moves we make in language-games have their power, so to speak, independently of whether we mean them. (If for devious reasons I tell my wife that I will quit tomorrow if she encourages me, and she does, I have as much commitment to quit as if I had meant it. And if she encouraged me *just because* she suspected I was not in earnest, and to call my bluff, when in fact I was in earnest, she could not complain if next day I did quit).

5 Because moves in the language-game may be hard to make, because there are things that, except for devious purposes, we can not make ourselves say, developments in the language-game will often show more clearly than anything else, and even in the face of other biographical evidence, what our attitude is, for example how seriously we have been thinking something.

6 There is thus a sense in which attitudes are *created* by developments in a language-game. If at some juncture in a discussion I find that I can usefully and seriously say 'I would quit tomorrow if you encouraged me,' that may really be my attitude. But was it my attitude prior to my having occasion so to express myself? That is a funny question.

The
concept
of
pain

I

1 This essay is about the concept of pain, and not about the phenomenon. I can perhaps only explain indirectly what I mean by the concept of anything. If a word is new to me but I can be apprized of its meaning by being given a synonym of it in my own or some other language, then even prior to my learning its meaning I 'had the concept.' If, however, all the synonyms of the word are also new to me, then I do not 'have the concept'; I have still to pick up somehow all the miscellaneous detail of the way this and similar words are used. In the case of the concept of pain I would have to learn, for example, that pains can be intense, mild, sharp, shooting, throbbing, but not interesting, pleasant, black, two inches thick or half an ounce in weight; that pains can be in the head or the shoulder, but not in the shoe or just to the left of the roses; that pains can come and go, but do not come from anywhere or go anywhere; that people can have diseases without knowing it, but can not have pains without knowing it; that if I have pain in my tooth and the tooth is removed, then although I no longer have the pain, the pain was not removed, but rather stopped. If I say that

It will be obvious how much, in writing this paper, I have been affected by Wittgenstein's discussions of the now-painful subject of pain. There will be less agreement as to whether what I say represents a correct interpretation of Wittgenstein. I believe that in fact it does; but I have made no effort to substantiate that claim. The subject of pain being an extraordinarily difficult one, it seemed to me better, rather than burdening the presentation with exegetical argument, to set out my interpretation as an independently arguable thesis, and leave the justification of it for another occasion.

the dentist removed my tooth and the pain and flushed them both down the drain, then I have not yet mastered the concept of pain; but if I say that when he removed the tooth, the pain stopped, then so far at least I do have the concept.

A concept is not an experiencable mental entity; but as to what it *is*, that is to say as to its psychological, biological, or metaphysical status, I question whether it is useful to try to say anything. It is sufficient if, whether or not it can be explained *why* this is so, there are certain ways of using a word that seem to us to be in order, others that seem wrong, and perhaps still others that leave us uncertain. Whenever there is extensive agreement in our acceptance or rejection of, or uncertainty about the way a word is used, it is useful to talk of *the* concept of that word. When there is not such agreement, or when we are confronted with strange uses confidently employed, it is useful to talk only of *my* or *your* or the Eskimo concept.

It is important that, on this basis, if there are some uncertainties about how to use a word, that will not necessarily show that there is insufficient agreement to warrant talking of 'the concept': it may be part of the concept that this or that usage is dubious but not unacceptable. If we are all uncertain whether pains can be interesting or half an inch thick, then we agree in that uncertainty. It is part of the concept; and anyone who ventures to affirm or deny the possibility of interesting or thick pains does not have our concept.

Moreover, in talking of 'the concept,' one is making no representations about anything that may be supposed to exist independently of human behaviour and development. One is not, for example, saying that such-and-such people must be in possession of the eternal concept, because they agree so solidly concerning it. Consequently, it will not be relevant to object that there is no word in some other language with the same set of uses as the word under discussion, or that the use of that word has changed since the seventeenth century. Those objections would only show that a concept was not universal and timeless; and we need not make any such claim for a concept.

In talking about concepts we are not trying to determine how things ultimately are. Our interest in concepts is restricted to what will serve a certain philosophical purpose. We suggest to each person what his concept of this or that is, with a view to showing that a philosophical problem has arisen through a misreading of the concept, and that the problem can be resolved by getting straight about the concept. For this purpose it is necessary only that any given person be persuaded that he has in fact misread his concept. The procedure therefore relies entirely on an appeal to each person to consider whether or not the suggestion we are making applies in his case. If it does apply, then his philosophical problem is so far solved, regardless of what he may know or believe other people's concept to be.

It is true that if each of us had a different concept, there would be little use putting out philosophical meditations of this kind for general consumption (although even then they might serve as a model that other people could *adapt* to their own case). But this and only this is the importance of any claims to generality that may be made or implied: the wider the actual generality, the greater the number of persons who may find the proceedings useful. If any given person does find the proceedings useful, however, it should be no concern of his if some other person, having a different concept, does not.

At the very worst, therefore, we are hopefully raising balloons in talking about concepts. We are purveying a medicine that perhaps only the odd person will find beneficial. But the *patient* need have no worry about this, since unlike some medicines, ours is neither expensive nor toxic. It is other *doctors* who may be concerned about the range of usefulness of the medicine. To them we can say that if we have done our work well, surely the medicine will be worth offering for general consumption, because there is in fact very broad agreement in the ways people use words. Most of our linguistic exchanges are quite successful, and we are not forever entangled in the confusions and breakdowns of communication that would be rife if everyone's concepts were different.

2 Why would we talk about the *concept* of pain, rather than about the phenomenon? Certainly not because there is no phenomenon to talk about. It is not, for example, like the supposed phenomenon of intending or of expecting. If we make the mistake of thinking of those words as the names of characteristic phenomena we will just for that reason have a problem of the form: intending is a phenomenon, but it is not this or this, and not that, so what ever can it be? We must talk about the concept of intending just because there is no phenomenon. But there is no question whether there is such a phenomenon as pain. I have one in my shoulder right now, and you will never convince me that I am only imagining it.

3 It is not that there is no phenomenon that we call 'pain,' but that there is little that is puzzling about the phenomenon.

It might, however, be suggested that there is a problem as to how to *describe* pains: for although pains may be sharp, leaden, shooting, such adjectives do not further describe *pain*, any more than 'dark red' further described *red*. Dark red is just another colour, and sharp pain another sort of pain.

If this is a problem, it is easily solved. To those who can see and have learned language, 'red' is a richly descriptive word, and stands in no need of further enrichment, and so also is the word 'pain' to those who can feel. There is no reason why there *should* be a way of explaining 'red' to the blind, or 'pain' to the insensate.

4 It is not just that the phenomenon of pain is not puzzling: philosophical questions about the phenomenon have a way of turning out to be disguised conceptual questions, or to presuppose answers to such questions. The phenomenological question, for example, whether pain is inherently disagreeable may look different from the conceptual question whether it is analytic that pain hurts, but the former question will turn out to depend on how we answer the latter. We might tackle the phenomenological question by trying to teach people to like pain, or by searching for people who did like it; yet if we found someone who gurgled ecstatically when burned with hot pokers or paid the dentist extra to prolong the drilling of a cavity, we would not know whether to say that what that person liked was *pain*. We would be strongly inclined to think that by some freak of nature this person experienced some other sensation where we experience pain. This inclination would be an expression of our disposition to treat it as analytic that pain hurts, that is, to say that whatever does not hurt *cannot* be pain. Whether or not we could go on from there to consider whether it *is* analytic that pain hurts, or whether we merely have a strong disposition so to treat it, will be one of the questions to the answering of which I hope this essay will contribute.

5 Suppose we were to concede the analyticity of 'pain hurts': might it still be argued that it is anyway *because of* the overwhelming nastiness of pain that this proposition is analytic, and that therefore conceptual questions depend on phenomenological questions?

If anyone argues that way, he is perhaps impressed by how hard it is to imagine anyone liking such a thing as *pain*. He thinks of a case in which he himself has been in pain, and says to himself that a thing like *that* just couldn't be enjoyed. But like *what*? Well, we focus on the sensation, and think of it as being inherently hideous. Here, however, the cards are stacked, because we are thinking precisely of a case in which we had a sensation and a strong negative reaction. We cannot however say whether it was that the sensation was so ghastly that we simply could not control our reaction to it, or whether we were just concurrently having a sensation and undergoing a paroxysm. What may make the pain seem so ghastly is the violence of the concurrent reaction; but if we could imagine having the sensation without the reaction, it is not so clear that the hideousness of the sensation would remain. Would we rush to take some aspirin just to obliterate the sensation, even if we had no inclination whatsoever to scream, moan, or double up?

If it is absurd to say 'I think I might quite like pain, if only I did not find myself so tensed up every time I had it,' it is absurd not because it is unthinkable that one should like anything so nasty as pain, but because it is unthinkable that we should call what we have when we are not 'tensed up' *pain*. Thus, again, it is a conceptual and not a phenomenological question we are asking.

6 If there is a connection between the analyticity of 'pain hurts' and the nastiness of pain, it is not established by noticing the nastiness and from it conclusing that we have no option but to withhold the designation 'pain' from sensations that are not disagreeable. If we treat a proposition as analytic, it is not because to do so corresponds to the facts, or because we have no choice. We do not *decide* to treat a proposition as analytic; we learn to do so in learning a language. If one were to decide for oneself whether to use a word the way other people do, the key consideration would be, not the nature of one's experience, but whether if a different linguistic policy were adopted one would be able to make oneself understood.

7 In sections 2 to 6, reasons have been suggested for undertaking a conceptual, rather than a phenomenological discussion of pain; but really we do not need reasons. We do not need to know that tennis is a better game than golf in order to play it. If tennis is a game, let's play it; and perhaps tomorrow we may play golf.

8 It is, to begin, a conceptual question how *exact* the concept of pain is. Could a competent English speaker ever be in doubt as to whether what he feels is a pain, rather than, for example, an ache? A learner of the language might be uncertain whether to call his present sensation a head pain; but is this either because there are headaches and also head-pains, and it takes training to distinguish between them, or because although there are not in fact head-pains, *there might have been*, but what we have in the head does not conform to the model, or satisfy the description of a *pain*, but rather

that of an ache? We talk of stomach-aches and also stomach-pains: do we know how to tell the difference between them, and might we sometimes be mistaken in calling what we have a pain, when properly it is an ache?

9 Those questions, I think, can be answered quite summarily. We have stomach-pains and do not have head-pains; but this is not because the phenomenological character of what we have in the head is strikingly or even subtly different from that of what we have in the stomach, but because the 'head-pain' usage has never come into currency. We have back-, stomach-, and head-aches, and not arm-, knee-, and hip-aches, but not because we do not have in the arm what we have in the stomach: we do say 'my arm aches,' and the fact that when that is truly affirmed I do not have an arm-ache is due only to the fact that the expression 'arm-ache' has no currency.

We are not, in short, guided in our choice of the words 'ache' and 'pain' by exact or even rough distinctions between sensations. Sometimes we can use either word; but where we cannot, it is not because we do not have what one describes and do have what the other describes, but because one of the words has no currency in that context.

10 More difficult is the question, to what does the word 'pain' refer? When I say that I have a pain in my shoulder just now, am I informing you of the location of a familiar kind of sensation?

That seems right enough on the face of it; and yet if there were a pain-reaction-killing[1] drug that when taken would leave the sensation unchanged but render the sufferer perfectly indifferent to it, I think we would certainly not say that the erstwhile sufferer was now *in pain*, and it would at

---

1 Here and in some other places I assume for the sake of argument that we can make a clear distinction between sensation and reaction in the case of pain. This assumption seems fair enough as long as we dwell on the fact that I see your reaction although I don't feel your pain, or on the fact that my back aches but the misery shows on my face, and not on the fact that when my back aches I feel miserable (all over?). What should we say? – that there are three things, the pain, the misery, and the look on my face, or that there is one complex thing with a sort of centre in my back? And if we take away the 'reaction,' as we are calling it, should we take away the misery, or leave it and just take away the miserable look on my face? If we do the latter, should we leave the kinæsthetic sensation (if there is one) of grimacing and just remove the grimace? These questions seem to me unanswerable, and therefore I do not think there is a clear distinction between sensation and reaction.

The idea of a pain-reaction-killing drug, it may be noted, does the same service as Wittgenstein's idea (PI, §288) of turning to stone but having the pain go on. But the twist that is here put on that remark is very different from the usual interpretation.

the very least be unclear whether we should say that he now *has a pain*.[2]
We say that a person is in pain when he has a sensation from which he
recoils; and we are simply not prepared for the case in which he has only
the sensation, without the least inclination to recoil. If a drug of this kind
existed and was used much, our language might shift in such a way that
doctors would ask people under the influence of the drug whether they
were feeling pain today, where it was, how intense it was, and they would
unselfconsciously report that, for example, the pain in their shoulder was
milder today.[3] But as language now stands, we would be less happy to call
these sensations 'pains,' and would cast around for some expression like
'pain-likenesses.'

11 There is a similar agrument (if it needs to be argued) that 'pain' could
not refer to the typical behaviour of the sufferer: if there were a drug that
killed the sensation without killing the reaction, or if there were a drug
that made people writhe and moan, without seeming to themselves to be
pretending, but without feeling anything, we would not I think say they
were *in pain* if we believed them when they said they felt nothing; and
again we would certainly not say that they *had a pain* in their finger if while
under the influence of the former of these drugs we pricked their finger
with a pin and they reacted in the usual way.

Moreover, we say that people have pains in their arms, but we could not
say, as we ought if 'pain' just meant 'reacting in a certain way,' that a
person had a reaction, or a disposition to react or whatever it would be,
in his arm. If there is any location at all of the disposition to moan, it is in
the throat, mouth, and face; but we would not say that a person had a pain

2   It is not a good objection, I think, that if there were no reaction to it we would not
    know where the pain was – that we know that the pain is in such and such a spot if
    we wince, for example, when the spot is touched. We are sometimes quite certain
    where a pain is although that spot is not at all tender; and we sometimes find that
    touching a certain spot gives us pain some place else.
3   It may be questioned whether the *intensity* of pain could be discriminated without a
    reaction to it: generally how much we react, or how hard it is to suppress a reaction,
    would appear to be the gauge of the intensity of pain. But pain to which we can
    scarcely suppress a reaction does have a different phenomenological character than pain
    about which we find it easy to be quite composed. It may be that people who all along
    had only had the sensation, never the reaction, would never learn what pain to call
    'intense.' But we *have* had the reaction, and *have* learned, and hence could distinguish
    mild from intense pain even without the reaction. For a time, at any rate. It is of course
    possible that, without our at first noticing it, the drug should also have the effect of
    inverting the 'intensity' of the sensation – making the greatest physical damage produce
    the 'mildest' sensation. But we could obviously discover this too.

there when pricked with a pin on the finger; and the best we could say would be that 'a pain in the arm' = 'a disposition to moan, etc. importantly connected with the arm' (in a way which we need not here work out: perhaps 'stemming from it' or 'focussing upon it'). But it is not on a location but on what is at the location that we focus when we complain of pain in the arm.

12 If pain is neither the sensation nor the behaviour, it may look as if it is some conjunct of sensation and behaviour. And here there would be at least two possibilities: (i) that 'pain' means sensation + behaviour – to say that a person is in pain is to say that he has a certain kind of sensation and that he is either reacting to it in a certain way or is suppressing a disposition so to react; or (ii) that pain is not both sensation and reaction, it is just sensation, but it is the sensation we have when we are disposed to react in a certain way. The sensation is here *defined* by the reaction.

The first of these views has the same handicap as has saying that pain is behaviour: a pain in the shoulder would be a sensation in the shoulder + a disposition to react occurring in the shoulder. But if the disposition to react is locatable anywhere, it is certainly not where the pain is located. Of course, no one would say that pain in the shoulder was sensation in the shoulder + reaction in the shoulder; yet anyone who held that pain is sensation plus reaction *simply*, could not disavow this implication. There is no way of giving a general definition of pain as having two constituents, such that when the pain is *located*, one of the constituents has that location but the other does not.

The second possibility, that pain is the sensation we have when we react in a certain way, is not so obviously wrong. It does justice to the old idea that pain is a sensation; it does not have the difficulty that other accounts have with the locatability of pain; it can cope with the fact that we would not call a sensation 'pain' if there was no disposition to recoil associated with it; and it is one of the things we might say to explain 'pain' to someone who did not understand the word: 'It is what you feel when you are inclined to cry out, moan, etc.'

This last point may also, however, be cause for concern about this account: 'It is what you have when you react in a certain way' is only one of the things we may say by way of explaining 'pain.' If someone did not understand that, we might alternatively say that it is what you have when you are kicked in the shin or jabbed with a pin; or again that it is what aspirin relieves. And if someone thought of misery or grief as what he feels when he moans, or of annoyance as what he feels when kicked in the shins, we might have to add quite a number of negative specifications, to the effect that it is *not* any of these things. In short the account, even with such elaborations of it as we have been suggesting, is not such as to lead one infallibly to *pain*. Any of these explanations *might* be successful, but they have to be seen as useful guides rather than philosophical definitions; and we are playing a fool's game if we try to weave into our account everything that we could usefully say to anyone who managed to misunderstand what we had said up to any given point.

13 There is an interesting consequence of the idea that pain is *what you have when* (you moan, wince, and so on, or when some part of your body is in a disordered condition): it leads to the idea that it is possible that *what he has when* is something which, if I had it, I would not call 'pain.'

The connection is as follows: we think that we are first led to the identification of pains by being apprized of the conditions under which we have them, and then having identified them, we thereafter call anything that is sufficiently similar a pain. But if, having got this far, I now call anything with certain now-familiar properties 'pain,' and if *what you have when* does not have those properties, then if I had the sensation that you have when you moan, I would not call it pain.

14 Something has gone wrong here, however. We said 'It is possible that what he has when ... is something which, if I had it, I would not call pain,' and the question is, when I 'have it,' is it necessary or is it not that I should 'have it when'? If it is what he has when you prick his finger, is it all right if I just have it, or must I have it when you prick my finger? If it is what he has when he writhes, is it all right if I have it without writhing, or must I also writhe when I have it? If I writhe when I have it, then I must call it pain just because I writhe; and if it is what I have when you prick my finger, then either it is pain or for some reason I do not feel pain when my finger is pricked.

We either use the 'what I have when' criterion or we don't. We must use it if we are to attach the word 'pain' to two things that are supposed to be different; but if we therefore do use it we are prevented from supposing that you and I might have the same sensation and one of us not call it 'pain.' If it is 'what we have when,' then we *must* call it pain. We try to avoid this by supposing that we *just have the sensation*, not under the circumstances that make us call it pain, but then the supposed difference between the sensations is no more interesting than the difference between my sensation of red and yours of blue. It is not interesting that two sensations are different, unless they satisfy criteria for being called by the same name.

15 Still one may feel that this argument relies on one way of putting the problem, which may merely be an unfortunate formulation. One wants to say: your finger is pricked and you recoil, and mine is pricked and I recoil, and each of us has something very definite and different from the pricking and recoiling when this happens. Can't this something be different in your case from what it is in mine? It is only one way of expressing this possible difference to say, as we said before, that if I had your sensation I wouldn't call it pain; but even if we can't express the difference in that way, or even in any other way, the idea of there being a difference here is surely very clear. We could no doubt never *find out* that there was this difference, since I can't have your experience and you can't have mine, and we can't describe them, except as pain. But can we not set it down as a truth about human beings that we can never know whether what another person feels when he is in pain is like or unlike what we ourselves feel?

16 The short way with the above view would be to claim, as a verificationist might, that the question whether what another person feels may be different from what I feel, being admittedly an unanswerable question, makes no sense. But to argue in that way is merely to stipulate that the question makes no sense: it would be better if we could *show* that it makes none. This will be attempted in the next several sections.

17 In general we do not know what to make of a question whether two things are the same or different, without some specification of the respects in which they might be so. If you ask whether my car is the same as yours, I will be perplexed until you tell me whether you mean same make, same colour, same idiosyncrasies, or what; and unless you can say *what* make, colour, idiosyncrasies, or whatever yours is or has. Moreover, these 'same x' designations can be expanded into various alternatives. If the question is whether it is the same colour or the same make, we can go on to develop examples: orange or mauve or ..., Plymouth or Ford or ..., and so on. But if you ask whether my *sensation* is the same as yours, what further explanation of what you mean will be possible?

It is important here to realize that there is in practice no difficulty as to whether two people's pains occur at the same time, or in the same place, or are similarly intense, or sharp or shooting or leaden. Our question is not as to *these* features, but as to the character of *what* it is that is for example now in the shoulder, and is intense and leaden (compare §3 above). We want to know whether the quality of the sensation itself is the same or different. Let us call this the 'tone quality.'

The question whether your pain is the same as mine, we may thus recast as 'Do your pains have the same tone quality?' But that will turn out to be no advance if we can still give no answer to such further questions as 'What tone quality has yours?' If our cars are the same colour, they are both red or both green or ...; but if our pains have the same tone quality, what might they both be? If we cannot say, can we regard ourselves as understanding the question whether they are the same?

18 It may, however, seem to us that we are in a position like that of a person who has noticed different colours, but has not yet learned or invented words to describe them. Like the person without colour words, we *have* noticed real differences, and it is just unfortunate that it is so difficult to develop a language to mark the differences. We perhaps think that we could, if we worked at it a little, develop a language of our own, but that the reason we do not do this is only that it would have so little use, since we could not explain to other people what differences our words were marking. We could not do this because we do not have available the usual methods of establishing a common notation – for example, pointing to a colour we both see and saying let's call that c, and then testing whether we understand one another (whether, for example, it is the *colour* pointed to that the other person is going to call c) by trying whether he will call all c things c and no non-things c.

So we may think that for our own purposes we would have no particular difficulty in establishing such a language, and that the difficulty of sharing this language is due entirely to the privacy of its objects. *We* know what we mean; we just can't explain to other people.

19 But now let us consider whether we could in fact make the discrimination ourselves.[4] Just as the surface of an object is a blend of colour,

4 What follows is a version of the so-called 'private language' argument, and certain contrasts between it and the argument that is often attributed to Wittgenstein might be noted. In the first place the privacy of the object plays a comparatively minor role, while what plays a major role is (a) the absence of any established pattern of distinctions to which a learner's ability could approximate and which would show whether he was making the same discriminations from case to case, and (b) such peculiarities of the object, having nothing to do with its privacy, as that it has no identity except in its attributes (the same pain cannot be now this and now that), and that its attributes are not separable from, for instance, its location, so that there is no way of 'ringing the changes' on the distinctions to see if one has got them right (one cannot subtract the fact that it is in the head from a headache). In the second place, the argument serves

texture, sheen, and we have to learn to distinguish colour-ranges from sheen-ranges and texture-ranges; so, we shall suppose, a pain has a fine blend of qualities and we have to learn to distinguish the x-range from the y-range, and so on. Could we do this?

In the case of colour-ranges we do it essentially by being drilled in it by people who have already mastered the system, and this I suppose involves at least two things: (i) varying the items that have a given colour in such a way as to guard against the student taking 'red' to mean 'made of paper' or 'shiny' or 'warm to the touch'; and (ii) that he should perform in a way satisfactory to the instructor. This second point seems obvious and hardly worth mentioning, but it is important, I think, for at least two reasons: (a) these complex discriminations are not so to speak guided by the nature of things, in such a way that if we forgot them they could be re-established just by having a good look around us, but are distinctions that *we* have developed for various purposes, and therefore it is our particular way of making them, our particular way of cutting the cake, that must be learned; and (b) knowing the system just is being trained, habituated to its miscellaneous detail. There is no other guide to getting it right than doing it the way a trained practitioner does.

In view of this necessity of conforming to a complex established practice, it is at the very least unclear whether a person could even establish his own *colour* language: whether there is any sense in which he could drill himself until he got it right, when there is no established practice to which his performance might approximate.

But, however that may be, there are special difficulties in the case of pain-discriminations: Could one make clear to oneself the aspect of pain (this is, what corresponds to its being the colour, rather than texture, sheen, shape, that a set of discriminations ranges over) that one was dealing with? Ordinarily it helps (a) to be able to say such things as 'No, not the shape, not the texture,' and (b) to be able to say either 'It is the respect in which these things differ,' indicating a range of items similar or identical except

the comparatively limited purpose of helping to show that it is not just another person that can do nothing with the question whether his pain is the same as mine: we do not understand the question ourselves. And in the third place, being an argument that reveals the peculiarity of what we want to say about pain, it only has the effect of isolating and eliminating certain queer questions, and does not entail any theories about pain language according to which we should no longer say, for example, 'My dear, I know just how it feels,' or according to which thinking that a person is in pain is not thinking that he feels something.

for their colour, or 'It is what these things share,' indicating a range of things with few or no similarities except their colour. But (a) we would clearly not initially have any words for the qualities of pain that we did *not* mean, so that verbal explanations would be of no avail; (b) it is difficult to see how we could do anything like saying it is what these (different) things share, or that it is the way in which these (similar) things differ. If we indicate the items in these groups as headaches or backaches, then (a) we will have no way of indicating the differences amongst various headaches or various backaches, and (b) pains as so described resemble one another and differ from one another in ways that we do *not* mean, for example in being steady or throbbing, localized or diffuse; and although we can say that it is the way they resemble or differ from one another *over and above* these resemblances and differences, we still have not found a way of making plain to ourselves or anyone else what form these further discriminations are supposed to take.

Moreover, objects may be individuated for the purpose of ringing the changes on colour discriminations because there are *things* which are coloured, and which may be moved about, arranged in different ways. But in the case of pain, what corresponds to the colour (the tone quality) we have treated, not as a property of something, but as the thing itself; and of course it is no help in individuating anything to say that it is what has the quality we mean. We may paint an object a different colour, but if we try to imagine a pain with a different tone quality, there is no longer any sense in which it is the same pain.

We can move a red book from one place to another without materially affecting whether it is that book or whether it is red; but can a headache be moved to the elbow? If I fancy I have something in the elbow that is quite like what I have in the head, how may I drill myself to see whether it is the tone quality itself I recognize, rather than the steadiness or the diffuseness, or the bothersomeness? Could a throbbing pain have the same tone quality as a steady one, or a sharp pain have the same quality as a diffuse one?

20 If the foregoing considerations persuaded us that we cannot make regular pain discriminations, even in our own case, we might still feel that there really are differences there: that the pain I have now has a certain quality, even if for all I know it is not the same quality as the pain I had yesterday in the same place and from the same causes. There is something definite there, which might be the same or different in your case. One knows the question is not a perfectly clear one; one may accept that neither

you nor I can answer it; and perhaps one is practical enough not to keep asking the question under such circumstances; but one may still think it may be set down as a truth about human beings that they can't know whether their pain is the same as another person's.

21 Yet if it is granted that we can make no systematic pain discriminations even in our own case, then the use of the word 'pain' cannot rely on such discriminations. An ordinary remark that one has a pain will therefore be making no representation as to tone quality; and the ordinary remark that you and I have the same pain will not be falsified by the supposition that yours has one tone quality and mine another.

22 A person might grant, though, that the tonal difference is no part of what we are concerned about in our ordinary dealings with pain, and also that we can't describe tonal differences, but still hold there are such differences, familiar to each of us, and that one can illustrate them to oneself. Pain in different parts of the body, he might say, tends to have a different tone; hence we can say 'Maybe there is some such difference between his pain and mine as there is between one of my pains and another.' This question, it would be said, does not presuppose that it is part of the concept of pain that it should have a certain quality. Moreover, it is a question we understand; we can illustrate the difference to ourselves, even if we can't explain it.

I think it would be a mistake at this juncture to take a line like the following: 'All right. But if you admit that you can't talk sense about it, then don't talk about it. Worry privately if you like about whether another person's pain might be like *this* rather than like *this* (thinking of different pains you have), but until you can explain what you mean by "like this" it will be useless to try and entangle anyone else in your question. Whereof one cannot speak, thereof one must be silent.' There is something incoherent about this. The person who says this either is or is not satisfied that there is a difference between one of my pains and another. If he is satisfied, then we ought to be able to proceed from there. But if he is not, then he should not even allow us to cherish our problem privately.

It would be better, I think, to say that although there may be differences, familiar to me, among my pains, they are still all *my* pains, and the recognition of the multiplicity of them only makes the question of the possible difference between mine and his more complicated to state. Where before it was as if there was one thing that we each call pain, and I was wondering whether mine and his might be different, now there are two or more. But I am not wondering whether his pain is like one sort of pain that I have and unlike the other sorts, but whether all of his are different from all of mine. And I cannot illustrate *this* possibility with the case of the difference between a headache and a stomach-ache, any more than I could illustrate the supposition that his colour sensations might be different from mine with the difference between my sensations of red and of blue.

I *might* of course be wondering something quite specific, such as whether his headaches are like my stomach-aches, and we have not said anything so far that would imply that *this* question would not make sense. But (a) the only reason for confining the question to such specific possibilities would be to find some version of it that did make sense. That was not the possibility that we began by entertaining. And (b) it is doubtful whether even this limited question makes sense. Are we to imagine him having a stomach-ache in the head? Or ourselves having a headache in the stomach? Of course one wants to say 'No, not a headache in the stomach, but the tone quality of a headache, the pain itself, only transferred to the stomach.' But do we know how to subtract the fact that it is in the head from a headache? And if we did that, and also subtracted the fact that it was in the stomach from a stomach-ache, are we sure that the remainders would be different? Are pains like colours, which can be disposed over different objects without significant alteration to their nature?

23 On the principle that if one cannot suggest even to oneself what kind of difference there might be between two things, there is just no way of posing the question whether they might be different, I have been arguing that we can not pose the question whether my pains might be different from yours. But now if this conclusion were granted would it follow that your pain is the same as mine, that is, that your *mild* pain is the same as my *mild* pain, your *shooting* pain the same as mine? I think this does not follow. In the same way that we cannot explicate the idea of their being different, we cannot explicate that of their being the same.

Then, after all, we just do not know whether they are the same or not? No, to say that is to slip back into assuming that we can make sense of the idea of their being the same or different.

24 The conclusion we have reached here seems sophisticated and counter-intuitive, and therefore one may well wonder whether it is in fact a general truth about the concept 'pain,' or only an end product of the 'what you have when' account of pain (§11): the best we can say given the question that arises from that account, whether my pain might be different from yours?

Yet is our conclusion sophisticated? Does it conflict with anything we do say about pain? When people say sympathetically to the sufferer 'My dear, I know just what it's like,' are they saying something that conflicts with the view that we can make no sense of the question whether your pain is different from mine, or equally of the assertion that it is the same? Do they falsely think that what they say makes sense? Do they merely believe or assume that they know what it is like?

No. When people say things like this they do not *mean* that the tone quality you are experiencing is the same as one they have often felt, but rather, for example, that they know how endless a minute can be under these circumstances, know the way one thinks one will go out of one's mind if it doesn't stop, know how the pains shoot, jab, and throb. In short, they know all the describable things about being in pain, including not only one's reactions to the pain, but the describable characteristics of the pain itself (whether it is dull or sharp, leaden or shooting). And, certainly, we do not just assume that we know what it is like in these regards. We can find out whether we know by comparing reports.

Perhaps it comes to this: that what we spoke of earlier as *what* it was that is intense, mild, jabbing, people can't compare. But that seems hardly to say that, for example, mothers do not know just what it is like to have labour pains. If we can make any sense at all of this supposed residual quality – that is, *what* is sharp, dull, and so on – it is not to *it* that they are reacting when they cry out in pain; it is not on account of *it* that they deserve sympathy, but on account of the way it shoots and jabs, or the way one feels one will go out of one's mind if it doesn't stop soon.

It is perhaps as if our pains had colours; and then some people said it was the colour they found unbearable, and wondered why it did not bother them when a book was that colour, or a car.

But *is* it what corresponds to the colour on this supposition that is so ghastly? I think no one would know what to say: and that at least shows that it is possible that it is not. But if not, then *what* is it? There really seems nothing else that it could be; and it is certainly not *nothing* that is so vexing.

Well, of course, it is *pain*. It is pain that is mild or intense, that jabs or throbs, and that we are inclined to say has a peculiar tone quality. But perhaps we have been thinking of it on the model of an insect pest, that buzzes and bites, but take away all that it does and you still have something substantial. Can we similarly take away everything describable from pain and have something left? I think not: it is not just that whereas we do have inert insects we never have inert pains, and therefore we have difficulty saying what the residue would be if we took away all the mischief in the case of pain – but if we had something painlike that did not *hurt* we would not call it pain. There is, that is to say, a conceptual impossibility about the subtraction in the case of pain.

25 What I have just been saying suggests the following diagnosis of the 'other pains' problem: that it arises from trying to think of pain as *a something*, as *that which* is intense, or throbs, or jabs. Having taken this step we then ask 'Well now, *what* is it that affects us in these ways?' and in our desperation to answer this question we say that it is a certain tone quality.

We are in difficulty, however, right from the start, because if there are tone qualities, is it not *pain* that has them? Is the tone quality any less a property of something than, for instance, the intensity, and are we not still left without an answer to the question what it is that has the quality?

We are confused by the grammatical similarity between 'the pain is intense' and 'the book is red.' The book has a wealth of identity-sustaining properties that enable us both to say *what* it is that is red, and to say that it is the same book when we paint it green; but replace intensity with mildness in a pain, and we will not know whether to say that it is the same pain.

(If one were to insist on attaching a sense to this use of 'same pain,' one might say that it was the same if it was in the same place and from the same cause. But here we would be constructing and not describing a concept: we no more employ this concept of sameness than, if the light coming in my window is fainter now, we have a use for the assertion that it is still the same light.)

Our troubles have arisen from a failure to see, or a refusal to accept, certain peculiarities in the use of the noun 'pain.'

26 What are these peculiarities? In the first place, the word 'pain,' although descriptive, does not denote something that is distinguishable from what is denoted by the adjectives that describe it. By this I mean that, for example, an intense pain is not like a red Chevrolet, but like a bright yellow: there is something that remains unchanged if the colour of the car is changed from red to blue, whereas there is no yellowness that remains unchanged if the bright yellow fades. The adjective 'bright' in 'bright yellow' consumes or fundamentally changes the meaning of 'yellow.' Bright yellow is another colour, and intense pain is another kind of pain.

We might say that the word 'pain' is descriptive, not by referring to something that can have properties, but by invoking a range or a pattern of adjectives or other specifications. The word 'pain,' that is to say, is like a signal that a certain range of questions is askable, such as where it is, what is causing it, whether it is intense or mild, whether it shoots, jabs, throbs, or is leaden. It also signals the appropriateness of a set of reactions, such as sympathy or admiration for the sufferer's courage.[5]

Secondly, I would suggest that it is characteristic of the word 'pain' that we scarcely know to what we should say the adjectives modifying it refer. I will say I have an intense pain when I am distracted and half out of my mind on account of it. The pain is perhaps in my shoulder, but the suffering and distraction is either nowhere or all over me: does the word 'intense' describe the pain itself, or the reaction? Grammatically, of course, there is no doubt; but as I have suggested (§5), if we ask ourselves whether we react so strongly because of the intensity of our distaste for the sensation, or whether we concurrently have a sensation and react, and call the sensation 'intense' because of our reaction to it, we do not know what to say. All we know is that given the sensation and the reaction, we say we have an intense pain.

27 What we have been doing is peeling off a complicated conceptual structure from the experience of pain; and if I have not sofar specifically made the point, let me now suggest that we bring this complex structure to our experience. We acquire the concept from other people when we learn to talk, and it determines the way we treat our experience.

From this it may appear that the concept of pain is quite independent of the phenomenon. I do not think that this follows. We need hold no view as to how it is that the concept has evolved in the way it has: it might in some way historically have been determined by the nature of the phenomenon. I would want to say, however, that in the case of no individual does his experience of pain determine his concept.

To many people this may not be believable. It may be felt that no one who had not experienced pain could possibly understand the concept of pain. Our next question, then, is whether that is true?

28 I think it is fairly clear that all the things we have said about the *concept* could be both learned and applied by a person who had not himself felt pain. He could learn to tell when people were in pain, and when they were simulating it; he could learn what questions to ask and what questions

---

5 Here I do not mean that sympathy, for example, is *required*, but that the range of reactions from sympathy to callous indifference is appropriate in the case of an attribution of pain in a way that it is not appropriate if we say that a person has a stomach. One can not (logically) be callously indifferent to another person's having a stomach.

not to ask about people's pains; and he could come to be just as concerned and sympathetic as any of us are when he believed that people were in pain.

It might be held, however, that although he could learn all these things, he would not *understand* them: it would puzzle him why people found pain so objectionable, and the whole business about pain might at times seem a curious obsessive ritual that people went through.

Yet does he not understand? How would we decide this?

If we think of understanding as giving oneself a correct illustration, then of course the person who does not feel pain will not understand. But we sufferers understand the word 'pain' when, not ourselves being in pain, we cannot illustrate it to ourselves. The most one could say, therefore, is that understanding is being able to, being in a position to illustrate. We do not know, however, whether the person who has the concept, but has never been in pain, is in that position. He might very well recognize pain when first it struck; and if that is so, then prior to his first pain, he *was* in a position to identify pains.

Certainly, I think, when this person finally did have a pain, it would make a difference to his understanding. He would no longer be puzzled as to why people fuss so about pain, and would no longer be inclined to think of concern about it as being possibly a curious obsessive ritual. But it would not be the case that he now had a rational foundation for his concept: it would not be the case, for example, that from the sensation itself he could see just why we are so dreadfully concerned about it. He would still not know whether it was the inherent ghastliness of the sensation for which we have such a profound distaste, or whether it is bodily damage that jointly causes both sensation and reaction. He would understand, not in the sense of having a stateable explanation, but in the sense in which we understand when we 'know just how it feels,' or when we have been through it all ourselves. That is not a kind of understanding that will put us in a position to derive a concept from an experience.

29 The following problem arises out of the various claims we have made as to the referent of the word 'pain.' It was argued that the referent was not the sensation we have when we are in pain, nor the moaning, wincing, or other pain-reaction; nor is it both the sensation and the reaction; and finally it is not the sensation just when there is a pain-reaction. If that exhausts the possibilities, it would follow that the word 'pain' has no referent.

It is important to understand that in saying this we would not be saying

that there is 'no such thing' as pain, but only that the relation between anything or everything that we experience when we are in pain, and the word
'pain' is not the peculiar philosophical relation of *referring*. (This is not to
say that we can never, using the word 'refer' in the ordinary way, ask or
answer questions like 'To which pain were you referring just now?' but
only that when we speak, for example, of 'the pain in this shoulder,' nothing
is settled as to what exactly we mean.)

The problem generated by this conclusion is that even if the arguments
supporting it seem persuasive, one will have difficulty in seeing how, if I
say that I have a pain in my shoulder now, it can be true that I am not
referring to something of a definite kind, with a definite location at a
definite time.

What I propose as a solution to this problem is, first, that we are misled
by the grammatical similarity of 'a pain in the shoulder' and, for example,
'a chesterfield in the bedroom' into thinking that 'pain' must refer, as
'chesterfield' does, to a definite object; and, secondly, that the relationship
between what we say and what is the case is not one of referring, but of
saying ... when ... – that is, of saying various things under various sets of
conditions. I say I have a pain when (in a sense to be explained later) I have
a sensation to which I react; but I am not saying that I have the sensation
and the reaction. What I say invokes or makes appropriate a set of responses
of the kind outlined in §26. Part II will be an attempt to explain more fully
the suggestion I have just outlined.

II

In this Part, with a view to providing a perspective on the conclusions so far
reached, I will discuss the question how we learn pain language. Some
aspects of this question seem to me to present no difficulty, and I will not
discuss them. It is obvious, I think, that we learn a great deal about pain
language just by listening to people talking. We learn in this way that
people have head-aches and back-aches, but not knee-aches or shoulder-
aches; that pains may be mild, intense, throbbing, but not red, half an inch
thick, or soluble in water; that pains may be in the shoulder or the back,
but not in the mind or in one's pocket; and so forth.

The question I do wish to discuss is how we learn when to say that we
are in pain: how the connections are established between pain language and
our experience, or how it comes about that we can use pain language in
suitable ways and on suitable occasions.

While I think it is rarely true that we are *taught* when to say we are in pain, I will for convenience employ the supposition that everything from which (possibly over a long period of time and by various means) we learn 'what pain is' might be compressed into a few lessons. Moreover, I will discuss the question, not how we might teach someone who could already use a good deal of sensation language but not the word 'pain,' 'what pain is' – but how we might instruct a person who had so far mastered *no* sensation language. The former enterprise might succeed in one blow, for example, by saying that pain is the frightful sensation you have when kicked in the shins; but we would not thereby explain how we come to understand the words 'frightful,' 'sensation,' and 'have' (as here used).

To instruct a person without using (though not without mentioning) sensation language, we might begin by kicking him in the shins and saying 'that is pain.' He could misunderstand this: he could take it that 'pain' means 'having someone kick you in the shins,' 'kicking someone in the shins,' 'nasty human behaviour,' 'annoyance,' 'indignation,' or even (if we could suppose he could have this thought without all the words) 'what I feel in the shins when kicked, but perhaps not what I feel in the finger when cut.' In various familiar ways, however, we could discover that he had misunderstood; and we could try to steer him right by a programme of blocking out misconceptions: it is not this and it is not that, but it is what is left over when one subtracts all the things that it is not. To this we might have to add that it is not just what is left over in the case when he is kicked in the shins, but in a large variety of other cases, which we would have to describe severally in some detail, and for only some of which we could give a live demonstration. (One of the things the learner would have to get straight about is that the word 'pain' may correctly be used whether or not there is a readily assignable cause.)

Suppose now that, whether right away or at length, the student does understand; what should we suppose has happened? Has he managed somehow to single out just the right feature of his experience, and has he done something tantamount to saying to himself 'This and just this is the pain'? Has he formulated, or done something tantamount to formulating, a rule to the effect that in future anything sufficiently like this he will call 'pain'? Has he decided what the similarity is between leaden and sharp pains, or throbbing and shooting pains, that entitles them all to be called 'pain'? Has he reached a decision as to the role to assign to his reactions, for example whether being in pain is having a sensation like this and reacting like this, or having a sensation like this but only if he reacts this way, or even whether the sensation need not be like this, but pain is *whatever* sensation he reacts to in this way?

We may be inclined to suppose that in the course of the student's lessons, he must have been sifting, sorting, guessing, formulating hypotheses and testing them, until he hits on the right answer, or gives himself the correct ostensive definition. I shall sometimes call this the analytical prowess theory. We are, I suggest, inclined further to suppose that only if the student has in some such way as this arrived at the correct answer to the question what pain is will he be able to use the word competently: the use of the word derives somehow from the nature of what it stands for.

Here, it is worth noting, there would be no problem of what pain is if we did not think that each of us knows what it is. We think that we all certainly know what pain is: that is what we have found out in the course of the learning process, and that is what explains our ability to use the word. We know what pain is, and therefore we should be able to *say* what it is. The problem is to devise a way of saying it, or a reason why we can't say; but if we do not (in the relevant sense) know, that is, if we have not settled on an answer to the question what the word stands for, it will be no problem why we cannot say.

I have described a certain account of how we learn language, and have suggested how it leads us into various difficulties. I think many people would say that these difficulties must simply be faced, since no other form of account is possible. Later I will suggest some criticisms of the analytical prowess theory itself, but at this point I would like to show that it is at least not true that there is no possible alternative. For this purpose it is not necessary for me to claim that my alternative theory is *true*, but only that it is a possible theory, and therefore we are not bound by the absence of alternatives to the analytical prowess account.

Let me then suggest two things: first, that the objective of the various sittings involved in learning to use a word is not to guess the formula, arrive at the correct account of what the word means, but simply to bring it about that the student will use the word whenever the teacher would use it. There need be no rationale behind the student's performance when he achieves an acceptable competence: he need not, for example, be deriving what he does from some rule that he has decided upon. He may only have acquired the habit of speaking in the ways we would speak and not speaking in the ways we would not speak.

Secondly, I want to suggest that if there is any sense in which we teach the student in what circumstances to use a word, then even if those circumstances include the having of some experience, we are not teaching him *what* circumstances these are, we are not for example teaching him various *lists* of circumstances, so that he may later decide whether to use the word by checking over his lists; but we are working with him in such a way that in appropriate circumstances he *will* use the word.

What I mean may perhaps be made clearer by the following example: it is as if there was a machine that went through various contortions, but was also so designed that if we pressed a button at any particular point in its history, that would bring it about that henceforth whenever the device returned to the state it was in at the moment when we pressed the button, it would do something, say, ring a bell. The machine has various bells, corresponding to the various words we might want to teach a person; and for the ringing of any one bell, we could press the button at various junctures, and so bring it about that for any one of quite a large number of positions, whenever the machine returned to that particular condition, this bell would ring. When we finished setting the machine up in this way, it would not be the case that it 'knew' something about when to ring that bell, that it had arrived at some decision as to what we were getting at, it would only be the case that the bell would ring on all the right occasions.

In this analogy our pressing the button would correspond to our saying in various circumstances 'Now you are in pain'; the various conditions of the machine would correspond to various circumstances in which a person would be said to be in pain; and the ringing of a certain bell by the machine would correspond to the learner's use of the word 'pain.'[6]

The machine model as so far described is, however, too crude in at least two important ways: (1) a person may, when kicked in the shins, for example, be both in pain and indignant; and we have provided no way of explaining how it is that it is just that part of the state of the machine that

6 The foregoing analogy might serve as an illustration of the following remarks from *Zettel*:

114 One learns the word 'think,' i.e. its use, under certain circumstances, which, however, one does not learn to describe.

115 But I *can teach* a person the use of the word! For a description of those circumstances is not needed for that.

116 I just teach him the word *under particular circumstances*.

119 If I have learned to carry out a particular activity in a particular room (putting the room in order, say) and am master of this technique, it does not follow that I must be ready to describe the arrangement of the room; even if I should at once notice, and could also describe, any alteration in it.

corresponds to *pain* that causes the 'pain' bell to ring. This suggests that some singling or sorting out of aspects of the machine's condition would be necessary, and thus we are threatened with a return to the analytical prowess model. (2) The model as so far described does not enable us to explain how it is that from a few examples we can learn how to use a word in *all kinds* of circumstances.

The former difficulty could perhaps be taken care of in either of two ways: (i) by supposing the machine to be wired in such a way that a signal may initially ring any one of several bells and we therefore have to proceed tentatively in our pressing of buttons, and if the wrong bell rings compensate with a number of further well-chosen button-pressings until we find that the tendency for that bell to ring is cancelled and the right tendency established. We might suppose two kinds of buttons, 'Try' and 'Fix' buttons, the former of which would set up tentative and erasible connections, and the latter strong or hard-to-erase connections. These would correspond to our saying, for example, 'Perhaps this will explain it,' and the latter to our saying 'Yes, you've got it.' Alternatively (ii) we might suppose that the relation between button pressed and bell rung was variable and depended on what kind of connections were already established with the other available bells. If there was already a strong pattern established for the 'indignation' bell, or if a good deal of circuitry for the 'pain' bell was already operational, a machine would be ever so much less likely to take an exhibition of pain to be one of what we should call 'indignation,' while if little or nothing were antecedently established, it would only be by chance that an exhibition of pain would result in the ringing of the 'pain' bell.

The difficulty as to how we could learn from a few examples to use a word in all kinds of cases might be handled first by supposing that it was part of the design of the machine, not just passively to register whatever connections resulted from our button-pressing, but (by whatever means) to anticipate what buttons we would press by ringing its bells in various circumstances not yet prescribed. It would sometimes get this right and we would press the 'Fix' button, and sometimes wrong, in which case we would institute the same sort of button-pressing programme that we would employ to correct any of its other mistakes. I think we should suppose further that there is not in principle any end to this state of affairs: it is not the case that the machine at some early or even late stage reaches the point where it will henceforth make no mistake about what bells it rings; but *every* time it rings a bell it stands to be corrected; every time we use a word it is possible that we are using it ineptly.

The 'analytical prowess' theory does provide a ready account of the two difficulties we have just been discussing. To the question how a person misunderstands an exhibition of pain it answers that he suggests to himself a plausible but mistaken hypothesis as to what we are getting at. To the question how a person comes to use a word correctly in ways not previously illustrated to him it answers that he has generalized from the examples he *has* seen, hit on what they have in common, and is following the prescription contained in that generalization.

The analytical prowess account has its own difficulties, however. They are of two kinds:

1 There is a host of empirical difficulties which I will not go into in detail, as to whether in fact we all do arrive at an understanding of words in the way this theory says we do; as to whether we can do it without being aware of it; as to why, if we have arrived at some rule or settled on some experience as our guide to the use of a word, we are not more ready with an account of what this is, and are prepared to give different accounts and to revise our accounts.

2 There is a particularly serious difficulty as to how, if we did have rules for the use of words, we would operate with them. When I say I have a pain in my side, according to the analytical prowess theory what has happened is that I have started with a sensation and, guided by some rule, derived the conclusion that the thing to say here is 'I have a pain in my side.' But now how does the rule refer to the sensation? It can not say 'When you have what we call "a pain in the side," say that you have a pain in the side,' because that would presuppose that I knew without the rule when to say that I had a pain in my side. Nor can it say 'When you have something in the side sufficiently like this sample, say that you have a pain in the side,' not only because the sample would itself have to be a pain, and then we could say we had one pain only when we had two, but also because this rule would presuppose that we could, without the rule, identify the sample as a pain. And any other alternative, such as 'When you have a frightful sensation, say you have a pain,' would merely be a rule for connecting the word 'pain' to some other word or expression, which itself could no better be connected to the sensation than can the word 'pain.'

(Our machine model, it should be noted, has no difficulty with this question. Where the analytical prowess model requires that pain should be identified but is logically incapable of providing any equipment for this, the machine model requires only that it should occur. Its occurrence (together with other things) *causes* us to say we have a pain.)

I would like to comment more particularly on what I think is taken to be an overwhelming consideration in favour of the view that we learn language by formulating hypotheses as to how words are used; namely, that this view seems the only way of accounting for the fact that we can use words correctly in contexts in which we have never before seen them used. What, it is argued, could explain this better than to say that if a person is right in the rule that he develops, then of course it will apply *anywhere*? In spite of the above objections one may still be extremely puzzled as to how this human ability is possible if we do not use rules.

I think some of the intensity of the puzzlement here may derive from the supposition that at some (quite early) point everything about the use of a word becomes clear, and thereafter we at least have all the equipment we need for proceeding correctly in every possible circumstance: we will generally use a word correctly; but if we misuse it, our mistake must be due, not to our not knowing the correct use, but to carelessness or bungling of some kind. This supposition is suggested by ways in which we often put the question, for example, 'How is it that we *know* how to use a word in new contexts?' And it is of course very difficult to explain this, if it means that we are equipped to be *right* in our use of a word in these new contexts.

But now do we in this sense *know* how to use a word in new contexts? Perhaps we just try it out in new places, and sometimes the venture is successful, other times not. Perhaps we say we have a knee-ache, and people smile and say they know what we mean, but that one doesn't call it a 'knee-ache.' And then perhaps we will not know whether to say that our knee aches, although in this case we would have done all right.

Surely something like this is the way it is. There are, of course, extremely large areas where there is nothing experimental about the way we express ourselves, but these are just the areas in which we have successfully used a word many times before. There are also areas where an *individual* may not know how to express himself although there is an established usage, and still further areas where there is not established usage today, although there may be next year or ten years hence. In the last-mentioned areas we will simply manage as best we can; and some people will and others will not think they know what we mean; and sometimes a way of speaking that we devise in some unusual circumstance will take root and become established other times not.

On this basis there will be no great difficulty explaining our ability to speak and to be understood in new circumstances. That is a problem only if we suppose that we know just exactly what to say, and that other people will of course understand us perfectly. But in novel circumstances we do not know just what to say; nor do other people have no difficulty with the words we may venture. We muddle through; and it is if anything only a routine task to explain how we do *that*.

Let me now review the contributions that the machine analogy makes to our discussion of the concept of pain.

First, as we have seen, it suggests that when we talk of using a word in certain circumstances, it will by no means be necessary that we know what these circumstances are, although, of course, with a rather more complex machine than we described, it would be possible that it should. But even then the machine would not *employ* this knowledge in ringing its bells, any more than people do in speaking. People do not, in speaking, employ lists of circumstances, or principles for choosing the relevant circumstances: but when the circumstances occur we as it were automatically say the word (or, since we do not always say we are in pain when we are, we will automatically be so disposed as to say it).

Secondly, since on this model there is no exercise of shrewdness and analytical skill in learning the use of a word, it will by no means be necessary either that *we* should have answers *or that there should be answers* to such questions as 'What is pain, exactly?' On the 'analytical prowess' view of language learning the correct use of a word would be derived from a correct answer to this question (or how nearly right anyone's use was would depend on how nearly right his idea of pain was); but on our machine model, no such guide is required, and the correct use of a word like 'pain' will not be the use that is consistent with what pain really is, but the use that accords with the linguistic habits of the race.

In this connection, and thirdly, it may be important to notice that it is not necessary that the tuners of the machine should know what pain is. We might easily suppose that we, the human button-pushers, would know, and that it is just the machine that never does. But in the analogy, of course, we are all machines together: the tuners are machines that have in turn been tuned, and they press the button whenever conditions like the conditions for their 'pain' bell ringing occur. Teacher and student do not work together by the teacher imparting to the student something that he knows but can only express indirectly, but the teacher tries to bring it about that the student's pain bell rings on all the same sorts of occasions that his own does. And for this the teacher does not need to know, be able to give a general account of, what they are, but only to run through a variety of them, finding out as he goes when his pain bell rings, and doing whatever he can to make the student's bell operate the same way. I might illustrate the difference I have in mind here with the following figures:

Figure A   Figure B

The wiggly lines in these figures represent the complex use of a word. T stands for teacher, and s for student. In Figure A the teacher sees the course to be followed and tries to get the student to follow it. In Figure B the dotted line represents the teacher's already established competence in the use of the word. This is not something that he 'sees,' but something that appears as he goes along, for example in his answers to various questions as to what he would and would not say. His aim is to train the student to go the way he goes, and this is represented by the parallel lines in Figure B; but since he does not 'see' the path ahead, he teaches, not by calling out directions from a vantage point, as in Figure A, but by going along the path with the student, finding the path as he goes.

Here I am of course not recommending a method of teaching. I am sure that, people being as they are, it will be useful to sketch general views of the course to be followed or point out similarities and differences in parts of the landscape. But I think we may conclude too hastily from the fact that such measures are sometimes useful that they are essential, and that what we do is to equip the student with a complete set of them, a set such that, just from the general things we have said, the student could take all the right turns along the way. And I want to suggest that the *general* things we say, (a) are only one method among many of establishing linguistic habits, and (b) rather than forming up into a complete system are left behind and no longer necessary once the habits are well established.

An important point here is that if neither the teacher nor the student knows what pain is *exactly*, then there is no answer to that question. There will, sure enough, be innumerable things one can say to explain to another

person what pain is, and different things to different people who misunderstand in different ways or differ in their abilities to understand
explanations; but neither any of these things nor all of them together tell
us what pain is in the sense of exactly isolating just the right phenomenon.
Most of us 'know what pain is,' but that ordinary expression does not mean
that either we are ready with a good answer to the question what it is, or
it is a deep philosophical problem why not. When the sufferer says he
knows what pain is, he is not on the point of revealing the solution to an
old philosophical problem.

Briefly, then, the way in which this account of language-learning applies
to our earlier troubles is this: in line with the analytical prowess model of
language-learning, we are inclined to think that the learning process has
served to establish a connection between the word 'pain' and a discriminable
(inner) object or range of objects, having a certain property (tone quality)
or range of properties. But if this is so, then when we suppose another person has a pain, we will be supposing that he 'has' one or other of that range
of objects, having one or other of that range of qualities. Since that range
is supposed to be what pain means, then if what he has should be different
from that, then whatever it is, it is not pain. But now it must strike us that
we never know just what a person is experiencing when we suppose him
to have a pain, and hence we are never justified in attributing pain to him,
nor do we know what he means when he says he has a pain.

We may therefore make a shift and say that pain is not a specific sensation
or type of sensation, it is whatever sensation a person has when he is
wounded or diseased and when he moans, winces, takes aspirin. There is
then the same difficulty about knowing what he experiences, but it now
seems academic, and not such as to prevent us from attributing pain to him.
Should what another person has be different from *what we have when*, so
long as it is *what he has when*, it is pain, that being what we are now supposing pain to be. But this still seems to leave 'pain' just the name of the tone
quality, *whatever it may be*. And if we say that the pain throbs or streaks in
crooked fingers up our back, it seems to drain the streaking and throbbing
of all importance, and to demand that we sympathize, not because of *it*,
but because something we know not what, is doing it. Similarly, if we say
that the pain is unbearable, it seems to require that it is the tone quality,
rather than the streaking or throbbing, or the way we feel we will go out
of our mind if it doesn't stop, that is unbearable. But we cannot say, even
in our own case, whether this is true. There is a complex of conditions and

it is frightful. But, we have argued, 'pain' is not the name of the complex or of any family of them: the complex is not, for example, in the shoulder, but the pain is.

These difficulties all seem to arise directly or indirectly from the supposition that if we understand the word 'pain' we ought to *know* what pain is: for example, that we must have a model of a pain, and follow a policy of calling anything sufficiently like it a pain; or that we must have a definition of pain, perhaps as 'what a person feels when he moans and cries out, or when his body is disordered in certain ways.' But the model of language-learning outlined in this Part shows that it is not necessary to suppose that a person who understands the word 'pain' knows what pain is; and if we do not make the supposition, the problems that it generates will not arise.

In saying that we do not know what pain is, I of course do not wish to imply either that it is something, but we just have not yet found out what, or that it is nothing. This way of speaking is adopted only in contradistinction to the thesis that we do know what pain is, in the sense that we have some kind of a criterion of pain.

It may be asked whether the old problem of how we know that another person is in pain is not just as troublesome given our account of pain language as it is otherwise. Do we infer the pain from the behaviour, and if so, how do we know that there is the same kind of connection between pain and behaviour in other people's case, as there is in our own?

I have two main points to make concerning this question. The first is that in the average case, for example in which a person is wounded and writhing, there is just no doubt that he is in pain, and hence no inference is called for; and in the case in which we suspect that a person is dissembling, we are deciding as much about the genuineness of the pain behaviour as about the existence of something frightful. If, with an impish smile, a person says that he is in pain, what we *see* is off-colour, is not quite part of the pain-pattern. We do not infer the non-existence of the sensation, but rather we hesitate to accept the smiling profession as a case of being in pain.

Secondly, it is not, on the theory of this Part, a correct analysis of the judgment that a person is in pain, that he has a frightful sensation of the kind one often has oneself. That and similar analyses create two insuperable difficulties: how we infer to the sensation from the behaviour, and how we establish that whatever sensation does exist has just the properties of a pain, that is, how we establish the similarity of it to what I have. But if 'pain' is not the name of a sensation, then the judgment that a person has a pain will not be a judgment just as to the existence of a sensation, and still less as to the existence of a certain *kind* of sensation.

The usual way of generating the problem of other pains draws a hard line around the sensation, separating it from the bodily disorders and the pain behaviour, and requires us to bridge the gulf between sensation and other things, and to regard ourselves as assembling circumstantial evidence just as to the existence of the sensation. The approach of this essay does not allow that gulf to arise, and therefore does not require that we bridge it.

In supposing that a person is in pain, we are of course supposing that he feels something. One is not in pain if one feels nothing; but equally one is not in pain if, whatever one feels, one is utterly indifferent to it. We treat these two propositions as analytic; and the latter of them shows that reactions are as much conditions of saying that we are in pain as are sensations.

If, having concluded that a person's behaviour is genuine, we decide that he is in pain, one of the things we will have taken to be genuine will be his professions that he is in pain. Accepting the genuineness of them *is* believing that he is in pain. There is *no room* for an inference from that to something else.

We are not making inferences, but rather milking the concept of pain when we suppose that a person who is in pain feels something, and that it is in some part of his body, and that he is not indifferent to it although he may be brave about it, and that it may be mild or intense, sharp or leaden. If, milking the concept, we ask the sufferer where his pain is and what it is like, and he tells us it is in his side and it throbs, we do not make an inference from this behaviour of his to something lying behind it, but we either believe him or do not believe him, and if the former, what we believe about him is just what he has said, that he has a throbbing pain in the side.

Let me return now to the question raised in Part I, §4, whether it is analytic that pain hurts. I suggested just now that we treat it as analytic that what a person feels is not a pain if he is utterly indifferent to it; but there may be a problem how one could say such things given the kind of account of pain language I have suggested in this Part. There can be analytic propositions in formal systems, but the setting up of dispositions to use the word pain in various circumstances is not such a system: how then can we talk of analyticity?

The answer I suggest is that strictly there are no analytic propositions about pain, but that there is something resembling analyticity in the strength of our disinclination to say, for example, that a person is in pain if he feels nothing. If in all the cases in which I am tuned to say that I am in pain there is an unpleasant sensation, then I will just not know what to make of the supposition that I might be in pain, but feel nothing. It is not that I will *know* that idea to be wrong. There will be nothing to which I can appeal to demonstrate the wrongness of it. Anything I say will be, not a demonstration, but only an enlargement upon my discomfiture: I will be pinning down or elaborating the discord between the supposition and my linguistic habits.

It is, however, a merit of the theory of this Part that it does enable us to say that a person is not in pain if he feels nothing, and thus preserve one of our most stubborn intuitions about pain. This would seem a trifling accomplishment, were it not that on so many of the accounts philosophers are apt to give, we can 'divide through' by what we feel: it drops out as irrelevant. And where pain is concerned, that is surely a *reductio ad absurdum*.

On
how
we
talk

Our ability to talk sometimes appears to need explaining. It can seem puzzling, not only that we can deploy language to say such an immense number of different things and that we can understand so many things people say to us, but that we are able to relate language to our current needs and purposes, that we are able to determine that the sentences we use and not some others express what we want to say. (How, for example, are we able to say 'I am going to do it, but not because you asked me'? Is there a recognizable state of a person that we call 'doing it because one is asked,' and do we observe and report that this state does not currently obtain?)

One kind of explanation that attracts us is that of supposing that we are born with or acquire or are partly born with and partly acquire a system of operation, undoubtedly of a very complex sort, containing principles governing not only what will be a well-formed sentence, and what will be a meaningful sentence, but the relations between such sentences and whatever it may be that recommends their use at particular junctures.

Many people, one suspects, believe that there must be such a system, but until recently few have made any concerted attempt to map out exactly how it might work. One person who has made such an attempt is Jerrold Katz, and without making any claim as to how representative his work is, I will quote as an example of what I mean some of the things he says in *The Philosophy of Language*[1] as to how, in broad outline, this system is conceived to work.

1 Harper and Row, New York, 1966

Roughly, linguistic communication consists in the production of some external, publicly observable, acoustic phenomenon whose phonetic and syntactic structure encodes a speaker's inner, private thoughts or ideas and the decoding of the phonetic and syntactic structure exhibited in such a physical phenomenon by other speakers in the form of an inner private experience of the same thoughts or ideas (p. 98).

To understand the ability of natural languages to serve as instruments for the communication of thoughts and ideas we must understand what it is that permits those who speak them consistently to connect the right sounds with the right meanings.

It is quite clear that, in some sense, one who knows a natural language tacitly knows a system of rules. This is the only assumption by which we can account for a speaker's impressive ability to use language creatively. Fluent speakers both produce and understand sentences that they have never previously encountered, and they can do this for indefinitely many such novel sentences. In the normal use of language, the production and comprehension of new sentences, created on the spot, is the rule rather than the exception (p. 100).

... it seems necessary to conclude that speakers of a natural language communicate with each other in their language because each possesses essentially the same system of rules. Communication can take place because a speaker encodes a message using the same linguistic rules that his hearer uses to decode it (p. 102).

Roughly, and somewhat metaphorically, we can say that something of the following sort goes on when successful communication takes place. The speaker, for reasons that are linguistically irrelevant, chooses some message he wants to convey to his listeners: some thought he wants them to receive or some command he wants to give them or some question he wants to ask. This message is encoded in the form of a phonetic representation of an utterance by means of the system of linguistic rules with which the speaker is equipped. This encoding then becomes a signal to the speaker's articulatory organs, and he vocalizes an utterance of the proper phonetic shape.[2] This is, in turn, picked up by the hearer's auditory organs.

---

2 There would perhaps be no need for this talk of encoding and decoding if Katz did not regard a person's saying something as being essentially the making of some noises. Only if these noises are treated as being in themselves unintelligible will it seem necessary either to suppose that something that we understand is translated into them when we talk, or that they in turn need to be translated by another person into something else,

The speech sounds that stimulate these organs are then converted into a neural signal from which a phonetic representation equivalent to the one into which the speaker encoded his message is obtained. This representation is decoded into a representation of the same message that the speaker originally chose to convey by the hearer's equivalent system of linguistic rules (p. 103).

    While I cite these views of Katz as one example of what I mean, I would like the discussion that follows to be construed as having bearings on any explanatory project of that general type. Among such projects I would include any manifestation of the disposition in philosophy and psychology to suppose that when we say what we remember, imagine, think, wish, hope, intend, something happens in us that provides the material for what we say, and then we do something, or something further happens, that results in our finding the words to say it. A picture occurs and we describe it; a feeling occurs and we express it; and the words 'describe' and 'express' are here thought of as referring to a process or a procedure that generates sentences out of psychological events. It has not always been thought that there was any great problem as to how we find words for what we want to say, or anyway that problem has been deferred to some later time; and the emphasis has historically been on getting clear as to what exactly exists prior to its being cast into words. But clearly the two questions are inseparable, or if it is not clear this paper may make it clear. The people whose primary problem is what it is to have something to say before one finds a way of saying it subscribe as much to the picture of communication that Katz outlines as do the people (like Katz) who address themselves primarily to the question how we get from there to the things we say.

to be understood. My main interest in Katz in this essay runs in other directions, but I will here indicate how this feature of his theory might be treated.

    We might compare the words we use in talking with the money we use in commerce. Money can seem worthless to us when we reflect that it is only paper, and compare it with the valuable things we can buy with it; and similarly words can seem meaningless when we reflect that they are only noises, and contrast those noises with the significance they have when they are used in the course of asking questions, giving orders, telling jokes. But the money currently in use in a country cannot be said to be worthless, nor are the words currently in use meaningless. And just as the value of one paper currency is explainable in terms of other paper currency, so the meanings of words are explained in other words. Moreover, although one may buy a pound of butter with this piece of paper, the butter is not the value of the paper; and, similarly, although that building is called a house, the building is not the meaning of the word 'house' (cf PI, §120).

For convenience only, I propose to treat any supposition, no matter how general or how specific, that there is a regular functioning process or procedure by which we get from something that happens in us to its linguistic expression as a supposition that competent speakers are equipped with a talking machine. My justification for doing so is that I think one assumption we are bound to make about any such system is that while at any point *we* may still be unclear as to its working, the system itself is self-sufficient: it includes all the equipment necessary for grinding out the things we say without prompting or guidance. There are no gaps in it and nothing indeterminate about its mode of operation. If it were not so it would lack explanatory power; but if it is so then it is like a machine.

By a talking machine I of course do not mean a machine that talks, but a machine that might be part of a human being, and perform for him the tasks of finding words to express what he wants to say, or of showing him what is expressed by the words other people use.

In the first part of the paper I will explore the workings of the idea of a talking machine, making seven main points, which I will number; and in the second part I will discuss some problems as to its explanatory capability.

I

1 It is not clear whether, but for the idea of a talking machine, our ability to speak would appear to *need* an explanation. If we did not both suppose that there must be a system by which we arrive at what we say, and find it difficult to suggest what that system might be, there would either be no problem, or no clearly stateable problem.

If we are struck, as Katz is, by the fact that 'fluent speakers both produce and understand sentences that they have never previously encountered, and they can do this for indefinitely many such novel sentences' (p. 100), and ask 'How do we do it?' the question cannot have its usual sense. Usually when we ask how something is done, we cannot do it ourselves, and want instruction in it; but in the case of the question how we talk, we do it ourselves, and therefore in one sense at least we know how to do it, and cannot be intending to ask for instruction.

If I were asked how I talk, one kind of answer I might give is to attempt a description of how I move my lips and tongue to form various sounds; but if the other person said he was not interested in that, but rather in how

I decide what to say, I might take him to be a poor conversationalist who found himself at a loss for things to say, and I might be able to offer him some useful hints. I might say that the great thing is to acquire a certain brazenness, and overcome one's diffidence as to whether another person will be interested in what one says. That and similar hints would supply another sense for the question how one does it; but not a sense that is of any philosophical interest.

If the question how we do it were explained as being the question how, when one has something to say, one constructs a sentence that adequately expresses it, a response that might be given is to say that the problem does not arise, because this never happens. It is never the case that there is something definite there, whose meaning is transparent to me, but which I do not know how to express, and could use advice about expressing.

Two things *like* this that do happen are:

i One sometimes has something to tell someone, but has difficulty putting it tactfully, or perhaps one has a complicated tale to tell which may easily be misunderstood prior to the full spelling out of its complexities. The problem here is not one of how to encode something, cast it into words: for it may be something that is already in words, something one has read, or that someone has told us. And we would solve such a problem, not by artful encoding, but in such ways as by prefacing what we are going to say with a warning as to the necessity of hearing it to the end before coming to any conclusions. There is no philosophical problem *here*.

ii One is sometimes asked a question, the answer to which one is confident one knows, but which is difficult to answer, perhaps because one is uncertain how much background will need to be explained, and where is the best place to begin. Here we have a case of knowing what to say, but not yet knowing how to say it, in which the knowledge of what to say will not be experienced as having a verbal character. (The experience, if any, will be one of confidence that one can answer.) The ways in which we deal with this kind of situation are well known: we ask the other person how much he knows, and work from there, or we simply start someplace, and if the other person's reaction shows that our approach is too advanced, or too elementary, we try again in a different way. We do not study our knowledge of what to say, and work it up into articulate sentences. *It* is generally not in evidence. If the question is about what happened, we know we can answer it because we were there and saw it happen; if it is about how something works, we know we can answer it if we have studied how such things work. But our knowledge of how they work is not something that we experience as we explain it.

We have not yet found a way in which the question how we do it – how, when we know what we want to say, we arrive at a way of saying it – arises, in any philosophically interesting form. This does not show that there is no such question; indeed, we have proceeded in such a way as to exclude the question *a priori*, and those are tactics to be avoided if possible. We have looked for contexts in which, outside of philosophy, the question could be asked and answered. Given such a context, we understand the question all right; but just because the question can be asked and answered outside of philosophy, it is philosophically uninteresting. The effect of our tactics is to generate pressure for a clear specification of the question; but there is at least the following reason of a general kind for doubting whether such specification is possible: in other contexts in which the question how something is done arises, such as about baking a cake or playing chess, the person who does not know how to do something *can* ask the question, and can understand the answer; but if one does not know how to talk, one can not ask how it is done, or understand the answer. We can ask the question only in limited contexts, such as 'How do some people manage to talk so wittily, or so tactfully?' But Katz does not want to know how we talk *in certain ways*, but how we talk *at all*.

The sense Katz gives to the question how we do it is: 'what is our system for doing it?' He assumes that we do have a system, but that it is as yet unclear to us what that system is. '[That he knows a system of rules],' he says (p. 100), 'is the only assumption by which we can account for a speaker's impressive ability to use language creatively.' His question thus is not whether we have a system, but exactly what system we have. He does not adopt the idea that we 'know a system of rules' in preference to other *types* of answer, but never doubts that this is the right sort of answer.

There is an answer that he rejects, but it turns out to be simply a more complicated talking machine, and hence to leave him committed *a priori* to the idea that some mechanism will provide the explanation. He rejects the idea that we learn all the sentences we ever use, and that there is a separate and unique connection established between each sentence and the private thought or idea that (he thinks) it expresses. But this is only the idea of an ever so much more complicated mechanism that instead of constructing sentences to express ideas, finds which one of a huge number of pre-formed sentences fits an idea. He is not opting for a system of doing it as compared with some other way, but only for one sort of system as compared with another.

If we knew we had a system for talking, but did not know what it was, then the question 'what is our system?' would certainly call for an answer; but we do not know that we have a system. All we know is that we can do it. The prior question whether we have a system therefore needs to be answered before the question what that system is can have any urgency. We have here an avenue of escape from Katz's problem that was not available to him; but we cannot take this route until we have examined the credibility of the supposition that we have a system. This we will do in Part II. In the meantime we can only play along with the idea of such a system.

2 Given that the question how we do it makes sense, and that the general form of answer is to be that we do it with a system of some kind (a talking machine), it is fairly clear that we are required to look for some inputs for the machine, however in detail we may conceive its design. There might be machines that were so designed as to generate well-formed and meaningful sentences, but that alone would not serve to explain how we talk. The machine must find or construct a suitable sentence for our current purposes, and to this end it needs to be geared to something in us, something that it can express, describe, report. In this way we see how, whether or not it is obvious that when we talk there is something discernible that our words describe, report, express, or stand for, we are bound just by the form of explanation we have adopted to look for something of that description, and if it is not plainly given, to suppose that it is covertly given.

We need not enlarge upon how often, in the history of philosophy, problems can be seen as having been generated by just this consideration. David Hume, for example (*Treatise*, Book I, Part I, Section III), to answer the question how we tell whether an experience is one of remembering or of imagining, was moved to say that 'the former faculty paints its objects in more distinct colours than any which are employed by the latter. When we remember any past event, the idea of it flows in upon the mind in a forcible manner; whereas, in the imagination, the perception is faint and languid ...' Similarly, in writing about belief, Hume said (Part III, Section VII) 'But when I would explain [the *manner* in which a belief presents itself], I scarce find any word that fully answers the case, but am obliged to have recourse to everyone's feeling, in order to give him a perfect notion of this operation of the mind. An idea assented to *feels* different from a fictitious idea, that the fancy alone presents to us: and this feeling I endeavour to explain by calling it a superior *force*, or *vivacity*, or *solidity*, or *firmness*, or *steadiness*.' Hume, it may be suggested, is offering the languidness of some ideas, forcibleness of entry of others, and the hard-to-describe feel of still others as inputs for a talking machine, determining it to say whether we are imagining, remembering, or believing. He does not trouble to consider whether every use of 'imagine,' 'remember,' or 'believe' has its roots in such experiences, because he is sure *a priori* that something of this kind must be the case.

3 Not only must there be something to serve as input, but it must quite clearly be pre-verbal. If whatever it is that what we say expresses were already in words, no machine would be necessary. We would only have to say aloud what we had said to ourselves, and for that the machine would not be necessary.

The machine might still be necessary to explain how we arrive at what we say *to ourselves*, but this again would leave us with something pre-verbal as the input.

One might avoid this consequence by supposing that our 'private thoughts and ideas' were always in some kind of crude language, and that the function of the machine was to render crudely expressed sentences into well-formed sentences. (That is, after all, the main function of the grammatical rules that we urge upon children and foreigners.) But not only is there no evidence that this is so, there could be no better reason for supposing that crude expressions of what we want to say occur primitively than that refined ones do, or that we immediately understand ill-formed sentences, while having difficulty with those that are well-formed.

4 Not only does the idea of a talking machine entail that there must be inputs and that they must be pre-verbal, it entails also that they must be what we might call content-laden – that is to say they must contain in themselves somehow the entire content of what we end up saying. It is not, for example, good enough to say that the feeling of tension that we may have when we want to say something but have not yet found a way of saying it is the input, unless we can show differences in such feelings and a method by which from those differences we derive the different things we say on different occasions. Nor will the rambling imagery that we sometimes have when we are trying to remember something serve as input, unless there is something about the images that shows, for instance, in

what sequence and over how long a time the events we are trying to remember occurred.

The requirement that inputs should be 'content-laden' may come close enough to being satisfied in the case in which we give a general-purpose description of something we are currently witnessing; but we will see later (p. 159) that there are problems even here; and in almost all other cases the requirement will make it extremely difficult to suggest what might serve as an input for what we say. What, for example, might have served as the input for the sentence I am now writing?

5 It would not be so clear that inputs must be 'content-laden' were it not for the further requirement that the mode of operation of the machine must be neutral, that is to say the machine must be so designed as to be able to generate in a perfectly regular way the appropriate output for every input, guided by nothing not contained either in its own design or in the specific input. We cannot suppose that it 'just knows' that the feeling I just had was the feeling of wanting to say what I am now saying. This is, in fact, another expression of the principle on which the talking machine analogy was justified (above, p. 150). To suppose that the machine performs any feats of magic along the way would be to include a mystery in the theory that is supposed to dissolve the mystery of how we are able to talk.

The temptation to say that the machine 'just knows' can be illustrated by some perhaps naïve reflections about remembering, or about imagining. We may find that we do experience something that might serve as input when we remember or imagine, but that unfortunately the pictures, impressions, or whatever they may be do not have just the charactersitics that you might expect from what we *say* as to what we remember or how we imagine something. The impressions (let's call them) perhaps repeat themselves, are faint, appear in a certain order, and run their course in a quarter of a minute; while what we say we remember contains no repetitions, is either not faint or not as faint as the impressions, unfolds in a different sequence, and spans ten minutes. How have these adjustments been made? If we suppose that the machine just knows what modifications to make, we may as well suppose that it just knows the whole story, and can manage without inputs.

[For my part, I believe that something *like* that is the truth of the matter: that there *are no* inputs, and that a person (not a system for constructing sentences) does, when he knows at all, 'just know' what to say, for example as to the sequence or duration of certain events, or the brightness of a colour. He does not reascertain these things from his current experience: he knows what to say because he was there; or if he is explaining something, he knows what to say because he has training and experience in that topic. But to take any such approach as this would not be to modify, but to abandon the talking machine project.]

6 Some of the above-illustrated problems about arranging and assessing inputs might conceivably be handled simply by adding design features to the machine: by supposing, for example, that there is a regular relation between the time it takes the 'impressions' to unfold and the time we represent the remembered events as having spanned, and hence providing that the machine will regularly multiply by, say, 40; or by providing the machine with a regular way of allowing for the faintness of memory impressions, or of discounting repetitions of the impressions. But here a further feature emerges, namely, that such design features must work not only for individual cases but generally. If the machine discounts repetitions of impressions, it must still be capable of generating the statement that something was repeated; if it is so designed as to deliver the sentence 'I saw a bright star' on the stimulus of a faint impression of a star, it must also be capable of delivering the sentence 'I saw a faint star,' and so forth. In other words, the machine must 'know' when to disregard something (repetition, faintness) and when not to; but this once again introduces the element of magic; and some ways of avoiding supposing that the machine performs feats of magic only require us to make that supposition in other areas.

7 We can extricate ourselves from the kind of difficulty just mentioned by making one or other of two further suppositions about inputs: either that the 'impressions' always behave in just the way that the special design features require, for example that memory impressions of bright stars are faint and of faint stars fainter still; or that there are some other characteristics of the impressions that serve as a signal to the machine whether to apply its special discounting rules. Since the first of these options is clearly simplistic and not borne out by any experience, we are in effect required to make suppositions of the second kind. It is for this reason that philosophers have so often toyed with such ideas as that there is a feeling of pastness about a memory impression but not about a daydream or an hallucination, that we can tell how long ago something happened by the vividness of the impression we now have of it, that there is a feeling of rightness about the repetition of an 'impression' in the case in which we say that something was repeated, and many other things. If the talking machine is 'neutral' –

does not 'know' anything – there *must* be such guides to its operation. It is not that we find it to be the case that there are such guides, but rather that reflection on our explanatory model shows us that there must be.

Are there not feelings of pastness? Yes, there are. Nostalgia is such a feeling. We have it when we are struck by the unrepeatability of something that is past. It is not by having it that we identify something as being past, but we first recognize that a time in our lives is past, and then (sometimes) feel nostalgic.

II

In the foregoing I believe I have shown, with a minimum of comment on the merits of the scheme, how the idea that there must be some system by which we arrive at the things we say begets an elaborate set of requirements as to what happens when we talk. I propose now to raise some questions by way of critical evaluation of this kind of explanatory programme.

Suppose, to begin, that we were successful in designing a machine that would in fact do whatever jobs we required of it, such as reliably generating sentences new to it which are grammatical and make sense. I wish to suggest that we could not, without further evidence of a different sort, conclude that it is by virtue of some analogue of the machine's design that people are able to generate new sentences that are grammatical and meaningful. We need evidence, not only that the task can be performed in the way the machine does it, but that it is in fact so performed. Perhaps ideally that evidence might be that, with or without the help of the explanation, we are able to see that we do in fact follow just such a procedure as is represented by the design of the machine. If the explanation told us that there are rules that are employed in a certain fashion, we might find that we are conscious of using just those rules in just that way. But if this never became clear, as I think I need not argue that it would not, and we were thence bound to suppose that we followed the rules tacitly, implicitly, or subconsciously, it might still be evidence that we did so if the rules sounded familiar to us when we saw them formulated; if we were a good deal quicker at understanding how to use them than we would be in the case of equally complicated rules that were new to us; if the rules formed part of the usual programme of linguistic instruction; or if we sometimes forgot them and had to refresh our minds about them, or sometimes misapplied them and had to be reminded of their correct application.

In the absence of such evidence, the most we would be able to say for any explanatory system that we might devise is that it is such that it would yield the same linguistic performances that people exhibit, and that it *might* represent the way people function in speaking. It would be as if I were presented with a sealed device that would perform certain complicated manœuvres, and to explain what was inside, that enabled it to do these things, I designed and built a machine that would do all the same things in the same fashion. My machine does it all with gears, levers, and pulleys, but I cannot conclude from this that the sealed machine has the same insides: it might be an electronic device, or it might have dear little men working its controls.

It might be said that there is no logical difference between alternative explanations here, as long as they generate just the behaviour that the sealed machine exhibits. But surely all that is true here is that there is no logical difference between adequate alternative explanations of the same type. If two shafts turn at the same speed and we suppose them connected by gears, we need not ask how many gears or how many cogs, as long as the numbers are such as to deliver the required result; and it is perhaps all the same whether we suppose them connected by gears or by a belt and pulleys. But if *every* adequate explanation were logically equivalent, then explaining would consist essentially in plotting the resultant behaviour; and this is not true. The behaviour as plotted is what is to be explained, not the explanation of it. If the sealed machine in our example is a record changer, I plot part of its behaviour if I note that it does not *drop* the pick-up onto the record, but lowers it at a certain speed. How it does this is what I have to explain; and obviously there is a large number of available explanations. Its speed of descent might be controlled by the speed at which air escapes from a cylinder, or it might be geared to the motor.

If adequate explanations are not always logically equivalent, independent evidence of the correctness of any given design of talking machine will be necessary; but in fact there seems little prospect of such corroboration. The explanation of how we talk will no doubt be quite complicated; but we are not aware of doing anything complicated when we talk. We cannot break some process down into steps or stages, and when we have misspoken in some way analyze what we have done to discover whether we have omitted a step or misapplied a rule. The rules that philosophers offer to explain how we do it are not like old friends to us; we do not upon having them propounded to us immediately realize that we use them a hundred times a day; nor do we show any significant facility at understanding their application. And they are not generally rules that are part of any standard programme of linguistic instruction.

It is perhaps for such reasons as this that Katz says (*The Philosophy of Language*, p. 100) that we 'tacitly know a system of rules'; and by this he must mean that we must be presumed to know such a system even if there is no direct evidence that we do. He says as much (p. 181) when he likens this kind of explanation to the supposition of microentities and micro-processes in the physical sciences. I shall have more to say later about this turn of the wheel.

A second question, similar to the above, is whether or not it is necessary that there should be evidence of the existence of the inputs, over and above the fact that an otherwise satisfactory model of explanation requires inputs of certain kinds? Here again I should want to say that in the absence of such evidence there would be no reason for regarding any given model as an explanation of how people manage to talk, rather than just as another design of device that (interestingly) would deliver the kind of linguistic performances that people exhibit.

I shall take it as not requiring much argument that we at least do not experience inputs of the kind required by any explanatory model of the talking machine type. As was suggested above (p.155), the case in which we are describing what we are currently witnessing seems to be one in which there is something to serve as an input, which is pre-verbal, contains at least everything that is represented in the output, and offers no pressing problems as to its sequence, duration, repetitions, or vividness. Even for such cases, however, there is a problem how we would account for the very different things we would be likely to say about the same episode on different occasions or to different people or when in different moods. This is not a problem for Katz, because all such questions as what one's audience will understand, or will be interested in, or what one is oneself moved to single out, are presumably among the things that are treated by him (p. 103) as 'linguistically irrelevant': they determine the choice of 'message,' but it is only with the message once chosen that the talking machine has to deal. That, however, effectively disqualifies the passing scene as a possible input: it is not it, but the 'message' we choose concerning it, that is the input. (Moreover, it is not clear that we 'choose a message' in any other sense than that, for example, we decide to say 'He glowered angrily and went on with his work'; and that sense of 'choosing a message' eliminates the need for 'encoding' it. But this may be an issue peculiar to Katz's approach.)

When we move on from descriptions of the passing scene to cases of saying what one remembers, imagines, believes, wants, or means, or cases of asking questions, giving orders, or explaining how to do something or why one did something, it will be difficult to maintain that when one speaks one is always aware of an experience having the properties required of an input, particularly the properties of being pre-verbal and yet containing in itself what emerges as output. *Imagery* is the phenomenon that comes nearest to satisfying this requirement, since it is pre-verbal, content-laden, and capable of being described by regular means; but I can want, and say I want, a large red apple without having a mental or physical picture of one; and if I do find myself having a picture of an apple, that will not itself tell me whether to say that I want one, that I like apples, that I am imagining an apple, or that I believe apples exist.

Let us not go into the tiresome question of whether there are other experiences, concurrent with such a picture that tell us whether we want, like, hate, or are thinking about the thing pictured, and whether the picture represents some specific individual or any individual of that type. These questions need only be raised to show us the absurd lengths to which we will have to go to make the total experience sufficiently content-laden to serve as an input.

A more interesting set of implications of the idea that the experience we are having contains guides as to what we should say is that there must be a very delicate art of discriminating fine differences between psychical states, an art which we must have learned somehow, in which some of us will probably be more competent than others, and about the application of which we will sometimes experience uncertainty, and sometimes make mistakes. If such an art existed, it might be expected that some of us would be inattentive in our exercise of it, so that sometimes an experience would occur and there would be a moment of puzzlement before we realized that it was the desire to ask a question rather than state a fact or give an order; and we might sometimes be quite seriously puzzled as to *what* question we wanted to ask, and sometimes surprised or even annoyed on discovering what it was. Moreover, since the business of talking is conceived in a talking machine scheme as being one of accurately expressing our psychical states, it ought everywhere to be recognized that all a person can do is deliver himself of whatever inputs happen along, and therefore no one should object if, in the middle of a conversation about the population explosion, someone is moved to say that the rain in Spain falls mainly on the plain.

Further, there ought to be some kind of common knowledge of inputs, such that when a person has given a faulty explanation of something he does understand, we could help him to see how he might have misread the input.

It would be an everyday occurrence in such a world for people to say, for example, 'I believe that I believe that, however perhaps I know it,' or 'I think I want to ask you what time it is, but maybe what I want to know is what day of the week it is,' and many similar things.

The burlesque character of such a picture of the way we talk may dramatize the absurdity of the talking machine form of explanation; and yet there are features of this picture that may not seem untrue. We are, for example, sometimes unsure exactly what question we want to ask, or what point we want to make, and we can be uncertain whether we believe something, or what exactly we were just now thinking.

The question, I suggest, is not whether such uncertainties exist, or whether they can be resolved, but *how* they are resolved. On a talking machine account, they exist either because of some indistinctness of the psychical state, or because of some breakdown in the machinery for 'encoding' that state: a failure to observe some feature that is clearly there, or a misidentification of it. The uncertainty is resolved by closer self-examination and more careful application of the encoding rules.

In fact, however, while we have no very business-like or methodical ways of resolving these uncertainties, they are at least not resolved in *that kind* of way. A certain way of putting a question that we consider strikes us as insufficiently incisive, and after some head-scratching a better way occurs to us and we say 'That's what I wanted to say.' But we did not have an experience of an incisive question, which unfortunately did not come out as incisive when cast into words. (What kind of an experience is *that*?) We may just in general like to ask incisive questions, and have some confidence that we can do so. And we do not arrive at the incisive question by close examination of ourselves, but perhaps by reviewing the course of the conversation we are engaged in, looking for important issues that have not yet been canvassed, or for ways in which we may have been misled by some of the things that have been said.

When faced with the fact that we are not usually conscious when we talk of anything that will pass as an input, we are sometimes inclined to suppose that the inputs somehow lie hidden: that everything goes by too quickly for us to see it ordinarily, or that the process is somehow shy of being seen.

Talk in the usual way and it is all there but one is too preoccupied to be able to see it; but try to have a look at it and one is immediately doing something else, so that of course the process behind talking is not there to be seen. These are possible explanations of the unavailability of inputs; but one would still like something further to establish a difference between their not being available and their not being there at all.

Strong and weak versions of the explanations of the unavailability of inputs are possible: on a strong version, it is impossible, and not merely difficult, to experience the inputs. Such a theory rules out the possibility of establishing that they are there.

One would do better, therefore, to argue that it is merely *difficult* to experience inputs. Let us take this view, and suppose that under some conditions (we need not specify what they are), something that will pass as an input can be experienced. It will then be necessary to establish whether what we experience under those conditions is something that we have generated to satisfy the demand that there should be inputs, or whether it is a naturally occurring case of an input, without the occurrence of which we would be lost for something to say. (It is, after all, fatally easy, in considering whether, for example, we describe things from memory by conjuring up a picture and working from that, to satisfy ourselves that we do by doing just this. But the most such an experiment could show is that it can be done that way: it does not show that we always or even generally do it that way. And it does not even clearly show that it can be done: we might in such a performance simply be concurrently describing and imagining. The images might be illustrations of what we were saying, rather than what we said being descriptions of the images.)

We might try to eliminate this possibility that in our experiment we are creating the evidence to order, by simply trying to remember clearly what generally does happen. Yet if we then do remember some occurrences of something that might pass as an input, it is still possible that not only in contrived experiments, but in the regular course of events, the experience that we want to call an input occurs as an illustration or an offshoot of what we are concurrently saying, and only because we are so saying, rather than being the source from which what we say is derived.

It may promise to show that this is not a difficulty, if we can think of cases in which we had no input-type experience, and were at a loss for something to say. But then it would not be clear whether we were waiting for a picture (or some other input) to describe: if after a time pictures and words come, there will be the same question as to whether the pictures occur as illustrations of the words, or the words describe the pictures.

It might be that this question would be answered if we could remember cases in which the picture seemed new or surprising to us, and in which we had to study it to find out what to say, and were conscious of being guided in what we said by what current examination of the picture revealed. It must be noted that such a case would itself be somewhat extraordinary. Something that we *remembered* would not, for example, seem new and surprising to us; and if we had to examine a current experience closely to determine what to say, we would have no reason to think that we were faithfully reporting something other than that experience that had happened. The most likely sort of candidate for a case of this description might be an experience that we had under the influence of some drug, and which for some purpose we tried to describe as accurately as possible. Such a case would show nothing however about regular cases of talking.

Perhaps we could waive that difficulty, though, because there is an objection of more general application. When we can produce a case that seems to satisfy the requirements of our model, we are apt to think that it and cases like it are the test or revealing cases: they show clearly what is always there but often hidden. But with these or any other special cases that may seem to satisfy all the requirements, we will still have to establish a difference between the supposition that they are the revealing cases, and the supposition that they are simply special cases, and that other cases are not merely apparently but really unlike them. We do tend to be fascinated by the special case and keep coming back to it; but have we any reason beyond the fact that our explanatory model requires it for thinking that the same thing lies hidden in the common run of cases as is revealed in the special case?

There is, I have been trying to show, a rich array of perplexities that confronts us when we try to suppose that we are conscious either of the system that we use in constructing what we say, or of the inputs for that system. In view of such difficulties, it is perhaps understandable that one should vacillate as to whether these things are experienced, or whether perhaps they exist in the form of events in the nervous system. Katz, for example, began by cheerfully talking about experienced private thoughts and ideas as the inputs (or the outputs, when the machinery is reversed and we are understanding what other people say), but ended as we saw by talking about unobserved microentities and microprocesses.

We must therefore consider whether a theory in terms of microentities and processes is an allowable variant of an explanation in terms of experienced inputs and the conscious operation of a system of rules.

There will be some quite important differences between the two kinds of explanation. For example:

1 Central to an explanation in terms of conscious states and activities is the idea of proceeding on the basis of discriminated differences between inputs; whereas it is not clear how we could regard ourselves as discriminating differences between neurological events, and still less clear how we could regard the nervous system itself as doing this. It would have to be, not the *noticing* of a difference, but the *occurrence* of a difference, that would result in our saying something. The differences between the neurological events that serve as inputs would *activate* the neurological counterpart of a system of rules, and thereby *cause* us to say this or that.

2 Whether or not we must do so, we have assumed, and it seems reasonable to assume, that *error* would be a factor in the operation of a conscious system in the same sort of way that it is a factor in other human activities: we could fail to notice things, or notice but misinterpret them; we could forget or otherwise omit steps in a procedure; we could misapply rules. (We have, it is true, assumed that when functioning as designed, the system is mechanical, that is, that if everything is done in the right way, the right results follow; but we have not assumed that people will in fact always perform in the way the system requires.) It is not clear, however, that we could regard misfunctioning of a neurological system in the same way. The nervous system does not fail to notice things, forget to take certain steps, or to make certain allowances, and does not misapply rules: it just misfunctions. We would not point out to the nervous system where it had gone wrong, use a quaint analogy to explain to it a point about which it needed to be clear, or urge upon it the importance of remembering this and making allowances for that.

3 While it is clear enough, at least in certain cases (mental pictures are the best example), how we can understand an experienced input – or when the machinery is reversed and we are 'decoding' what someone else says, an experienced output – it is not at all clear how we would understand a state of the nervous system; or for that matter, how we would even know that it existed. It might be true that there is a state of the nervous system that prevails when we understand this, and another that prevails when we understand that; but to show that this was true, understanding would have to be defined independently of the nervous system; and it would just be an

interesting fact that when understanding as so defined prevailed, we could infer something as to the condition of the nervous system. And if neurologists discovered that there was no condition of the nervous system characteristic of understanding, that would not show that no one ever understood anything.

The foregoing considerations make it initially doubtful whether a neurological explanation could be of the same logical type as an explanation in terms of experienced inputs and the conscious application of a system. It might be replied, however, that there surely are neurological correlates of experiences, and of consciously employing a system of rules. A being without a nervous system or its equivalent could neither have experiences nor perform the operations that we call employing a system of rules. So why can we not suppose that even without its phenomenological counterpart, the neurological equivalent of experiencing things and applying rules to them exists in our nervous system when we talk?

It is difficult to deny that something happens in our nervous system when we talk, and there is no reason to suppose that we could not some day isolate the parts of what goes on that have to do just with arriving at what words to utter, from the parts that have to do, for example, with the physical process of articulating sounds or with our own reaction to the things we are saying, and so develop a neurological explanation of talking.

This itself, however, would not settle the question of whether the neural processes we had mapped out were the counterpart of using a system of rules, and it is difficult to see how that question could be settled if there was nothing with which to correlate the neural processes.

If we knew that the human nervous system was a rule-following device, then we would know that what we had mapped out somehow represented *some* system of rules, and it would perhaps be unimportant whether it was just the system we had contrived from an analysis of the outputs. Any two systems that were both rule systems and delivered the same output could be taken to be isomorphic.

But if we did not know the nervous system to be a device whose only mode of operation was rule-following, this argument would not be available to us. And this is what we do not know about the nervous system. It is not merely that it is not yet proven that it is a rule-following device: it is not clear that the idea of a device, rather than a person, following rules, makes sense; and if it does make sense, there seems every reason to suppose that the nervous system is not such a device.

One may think that there is no difficulty about the idea of a device following rules: it is just something that does what a person does in following a rule. A clear case of that is reminding oneself of the rule when the conditions for its application arise, and then applying it carefully in the way it directs. Hence, any device so designed that when certain conditions occur a mechanism is activated that makes it behave in a particular way is a rule-following device. Such a device operates in a certain way; but why should we say it follows a rule? It neither remembers nor forgets the rule, neither understands nor misunderstands its application, neither applies it carefully nor carelessly. It is not even clear what part of the mechanism is the rule. If I want a machine to follow the rule 'Whenever one revolution of this wheel is completed, turn that wheel through 10 degrees,' and I rig it with a tripping device, then the machine will perform as the rule prescribes, but the rule itself is no part of its design.

If we nevertheless assume that the idea of a rule-following device makes sense, and assume further that our behaviour is somehow governed by our nervous system, it is still scarcely believable that the nervous system (and hence our behaviour) is entirely rule-governed. Rules are in general of limited application: it is some feature or features of our behaviour that may be in line with some rule. The nervous system, we may suppose, controls everything we do; but the rules of any activity in which we are engaged grossly under-determine our performance. In playing golf there are rules of the game that we try to observe, and also rule-like injunctions that we bear in mind and try to apply, such as 'Keep your eye on the ball,' 'Keep your left arm straight'; but there are neither rules nor anything rule-like determining every flex of every muscle in a well-styled human movement. In fact, there are not rules governing any flex of any muscle. If a rule tells us that we must use a putter on the green, we must flex some muscles in order to follow it; but no rule tells us what muscles to flex.

It may, of course, be true that when a human being is taught to follow a rule, some change in his nervous system has been brought about of such a kind as to make possible what he then can do. I do not know whether this assumption is warranted or not; but if it is it does not follow that at least to that extent or for those cases the human nervous system has been made into a rule-following device. A person follows a rule; but the rule is not followed in miniature prior to or alongside his doing so. The nervous system cannot keep its eye on the ball, or repeat the same sequence of digits in the tens as in the units. And if it is nevertheless true that when a person follows a rule, there is something happening in his nervous system which occurs only at such times, this characteristic activity may be supposed to occur only in the cases in which, in the ordinary sense, a person follows a rule. But in the case of talking, it is just the lack of evidence that in the ordinary sense we operate a system of rules that leads us to posit micro-processes. We lack precisely the evidence we need to establish that rule-following is going on in the nervous system.

It should now be clear that positing microentities and processes does not simply amount to translating theories of the conscious operation of systematic procedures for talking into neurological terms, and is no way of saving those theories from the enormous objections to which they are subject. Not only can we not establish whether a microprocess is the counterpart of a conscious process when the latter does not occur; not only is it scarcely intelligible that the nervous system should notice or not notice differences between its states or should remember and apply rules govern-ing what to say, given those differences; the neurological explanation turns out to be of logically different kind; it does not tell us what understanding or knowing what one wants to say consists of, but tells us what should be found to be the case when, for other reasons, we are said to understand or to know what we want to say.

Let me conclude with a declaration as to where I think this discussion leads. We began by noting the way it may seem remarkable to us that we are able to speak, and that there seemed a very great problem of explaining this ability. It was first suggested that it was not clear whether the problem did not just arise because we made certain assumptions as to what form the explanation must take; and to this suggestion we might now add that in a way it is not at all remarkable that a human being should be able to speak. We have been doing it for ages; and it does not call for an explanation the way the sudden development of, for instance, the ability to walk on water might call for an explanation. It does not conflict with what we have all along known or supposed about human beings. We could also remind our-selves that anything can be made to look remarkable – the phenomenon of gravity or of momentum, for example; and most such things seem no less remarkable when we understand the usual scientific explanation of them. It is, therefore, by no means clear that the sense of amazement that we experience when we look at things in a certain way is a sign that they need to be explained.

While I mention these considerations, however, I am by no means satisfied that they show that there is no problem here. In spite of them I find myself craving an explanation of how it is that we are able to talk. What I think I have shown about such cravings is:

1 That it is by no means clear what kind of explanation we should expect.

2 It is extremely doubtful whether any explanation of the talking machine type will prove successful.

3 While I would reject any neurological explanation that was conceived to be the counterpart of a talking machine account, I do not believe myself to have shown anything as to the prospects of success of what one might call a plain neurological explanation, that is to say an explanation that (a) isolated the neural processes connected with our arriving as what we say from any other such concurrent neural processes as those connected with the physical process of articulating the words, with our own reaction to what we are saying, or with our hearing our own words spoken; and (b) simply described what was thereby isolated, without attempting to map it on to any such picture as that of choosing a message or encoding it. I am nevertheless extremely skeptical about such a scheme, for at least the following reasons:

i I cannot see what reasons we would have for saying that any process that we isolated in the above way was the process of 'arriving at what we say,' or of anything else if, *ex hypothesi*, it was not shown to be correlated with a process, otherwise discernible, which could be described as, for example, 'arriving at what we say.'

ii In any case I find it impossible to conjecture what explanatory force such a discovery would have. It would not, for example, relieve our puzzlement about 'the impressive ability of human beings to use language creatively' in the elegant way that Katzian hypothesis, if we accepted it, would. (It is not like being puzzled as to how someone does something apparently very difficult, and then being shown his system, and finding it easy after all.) It would show that we do have certain physical equipment for doing it; *but whether that is the case was never part of the problem.*

4 If the argument of this essay is sound, linguistic theorists are making a fundamental mistake if they suppose that generative grammar explains how we actually talk. This is not to say that such linguistic theories may not be useful human inventions. They may, for example, make it possible to

programme a computer to do translations; or they might form part of a technique for cracking codes.

If a planet follows a path around the sun that can be reproduced on paper by fastening a thread at two points and with a pencil describing the path that keeps the thread taut at all times, it does not follow that it is by such means that the planet's path is determined; and no more does it follow, if our linguistic output can be shown to have just the shape that would be generated by following a certain set of rules, that it is by our following those rules that it comes to have that shape. But just as, though not explanatory, the discovery as to the path followed by the planet is useful, so the discovery of a system for plotting the paths that language follows may, without being explanatory, be useful.

This distinction between the useful and the explanatory might, among other things, save us from having to say, with Chomsky,[3] such incredible things as that 'A theory of linguistic structure that aims for explanatory adequacy incorporates an account of linguistic universals, and it attributes tacit knowledge of these universals to the child. It proposes, then, that the child approaches the data with the presumption that they are drawn from a language of a certain antecedently well-defined type, his problem being to determine which of the (humanly) possible languages is that of the community in which he is placed. Language learning would be impossible unless this were the case.'

5 I think I may indirectly have brought out a point that may be important for further study of these questions, namely, that it is a mistake to suppose a sharp distinction between what we say and how we say it: to suppose, for example, that we first have something to say and then find a way of saying it, or that we first hear what other people say and then understand what it means. I suggest that if there is an explanation of our ability to talk, it will be one that, instead of treating language as a strange thing, a kind of code into which we cast our thoughts, explains how, for a being that is intelligent, alive, and a member of a species with a long history of language use, thought and its expression may make their appearance together; saying something can be our first reaction.

3 Noam Chomsky, *Aspects of the Theory of Syntax*, MIT Press, Cambridge, Mass., 1965, p. 27

Logical
compulsion

There is a somewhat untidy nest of related problems that can with some justification be called problems about logical compulsion, that is to say problems as to the nature or the basis of the incumbency that is upon us in formal (mathematical and logical) contexts to perform in very definite ways: to accept a certain conclusion if we accept certain premises, to get a certain sum when we add a column of figures.

The general approach to these problems taken in this essay is to attempt to show them to be confusions, arising in various ways from pictures or conceptions that we are deeply inclined to entertain about the nature of formal proceedings; and since there is no reason to expect that confusions will have an orderly form, it will not be possible to provide a succinct and comprehensive initial statement of the problem. Rather, the contours of the problem will have to emerge as we thread a more or less consecutive path through a somewhat complex maze.

Some things may usefully be said at this stage however by way of orientation. For instance, it may be important to say that although it will be argued that problems about logical compulsion arise from misunderstandings, the problems are not for that reason trivial or silly. These misunderstandings are not simple-minded blunders, like the failure to appreciate the ambiguity of a word or the failure to check whether some elementary empirical proposition is true. I am inclined to call them *deep* misunderstandings. They press upon us naturally and strongly; they are difficult to expose; we tend to slip back into them without noticing that we are doing so; and it is often peculiarly hard to see how things are if they are not as we picture them while under the influence of these misunderstandings.

The reason that the misunderstandings we are trying to expose are (in the above sense) 'deep,' it will be argued, is that there is a kind of fiction or conceit involved in the practice of logic and mathematics, without which those disciplines would not be what they are, and which in this sense is essential to them. The conceit must be taken seriously *within* formal proceedings. It cannot be rejected or abandoned; and this gives it an immense power to bewitch. We fail to understand its *status*, so to speak. There are various ways of expressing the conceit which, as explanatory metaphors, are perfectly satisfactory; and being satisfactory ways of explaining something which cannot be rejected, they make a powerful claim to be accepted as *true*.

The business of the essay will consist largely in attempts to confine various expressions of this conceit to a proper metaphorical status, without at the same time prejudicing their right to be taken seriously within formal proceedings. That proves to be a peculiarly difficult task. As we proceed, the hardest thing will often be to see how things are in logical matters if they are not as we fancy them when under the spell of the conceits that are essential to the proper conduct of such affairs. One may often feel that the conclusions suggested must be too radical, because it would be the ruination of formal proceedings to accept them. It may relieve such anxieties if two points are borne in mind:

1 The problems about logical compulsion do not arise while we are good-heartedly engaged in logical or mathematical activities. There is no juncture at which the correct solution to a logical or mathematical problem awaits the solution to a problem about logical compulsion. Nothing we say therefore can imperil formal proceedings.

2 The way things are in these matters is not generally that we are certain that we *have* the right answer, but only that we are certain that *there is* a right answer; and not generally that we proceed with calm assurance to work out the answer in a mechanically efficient way, but that we struggle to think what techniques to employ, make mistakes both as to what steps to take and as to how exactly to take them, double check our answers and apply to other people for a further check, and are frequently quite ready to

believe we are wrong. Urgency and anxiety, rather than calm and assurance, prevail. We muddle through.

Why should such homely facts as this be surprising? Because it is difficult to see how, things being thus, we can attribute to our procedures and results the certainty and exactness that we regard as their defining characteristics. Yet it is quite to be expected that fallible human beings should have to struggle and fret to achieve this exactness; and we do often enough achieve it, or at least we have as much assurance of having done so as can be gained from having carefully applied the usual tests and checks. That assurance is not equivalent to certainty; but only bewitchment by the ideal of exactness would make us demand greater assurance than is warranted when fallible human beings have proceeded with care and caution.

Many of the views expressed in this essay have come to me as a result of pondering remarks of Wittgenstein's in *Philosophical Investigations*, *Remarks on the Foundations of Mathematics*, and *Zettel*, and in recognition of my debt to him I will often refer to passages in these works that can, not always without the exercise of some ingenuity, be seen as expressing a view similar to mine. While the essay is presented as an independent treatment of logical compulsion, and while I say quite a few things that I do not find in Wittgenstein, I believe my discourse may be useful to Wittgenstein scholars in such ways as by defining a proper context for points, the place of which in an overall picture Wittgenstein often leaves unclear; by emphasizing points that may have been unnoticed or under-rated; and by suggesting a sense for Wittgensteinian remarks that may have seemed inscrutable.

While I will not argue the point, it may be useful if I suggest what seems to me the broad difference between my approach and that of most recent writers[1] on Wittgenstein and logical compulsion. In general, I believe, they attribute to Wittgenstein *answers* to questions in this area: they try to decide, for example, whether he is a conventionalist or a constructivist, or

---

1 In particular, Jonathan Bennett, 'On Being Forced to a Conclusion,' *Proceedings of the Aristotelian Society, Supplementary Volume*, xxxv, 1961, 15–31; Charles Chihara, 'Wittgenstein and Logical Compulsion,' *Analysis*, xxi, 1960–1, 136–40; Joseph Cowan, 'Wittgenstein's Philosophy of Logic,' *Philosophical Review*, lxx, 1961, 362–75; Michael Dummett, 'Wittgenstein's Philosophy of Mathematics,' *Philosophical Review*, lxviii, 1959, 324–48; E.J. Hall, 'The Hardness of the Logical "Must,"' *Analysis*, xxi, 1960–1, 68–72; Barry Stroud, 'Wittgenstein and Logical Necessity,' *Philosophical Review*, lxxiv, 1965, 504–18; O.P. Wood, 'On Being Forced to a Conclusion,' *Proceedings of the Aristotelian Society, Supplementary Volume*, xxxv, 1961, 35–44

exactly what version of such a theory he adopts; and in their critical delib-
erations they try to decide whether the answer attributed to him, or some
different answer is to be preferred. There is a radical difference between
any such treatment and the approach I have adopted, which treats the
problems as calling, not for solutions but for dissolution. Whether or not
in this I am right about Wittgenstein, I believe that so treated, the problems
about logical compulsion come out as being not only more interesting but
immensely more manageable.

The essay is divided into numbered sections, not only for convenience of
reference, but to mark shifts from one line of thought to another.

1 In the course of formal proceedings, we quite often use what we may
call words of necessitation, like 'must,' 'have to,' 'can only,' and 'cannot.'
We say that a person must draw a certain conclusion, that he cannot draw
any other conclusion, and that in a certain calculation he can only get this
result. We seem to regard ourselves as being compelled to perform in certain definite ways. Our first question, then, is what kind of compulsion is
this?

It is clear enough that, given two premises, we are not psychologically
or causally driven to a conclusion. If you propound the beginning of an
argument to me, I may fall asleep, wonder what you are up to, go for a
swim, or think about something else. Even if you *ask* me to draw the conclusion, I may fail to do so. I can as much think of something else or go for
a swim when asked to infer, as when left to think of it for myself.

And if I do adopt the project of inferring, the *correct* inference does not
press itself upon me, does not intrude itself with the persistence and tena-
ciousness of a cold coming on. If it is an unpleasant conclusion, I may sometimes feel some resistance to it, but it is not clear that at such times I am
fighting it but unable to repel its overwhelming force, rather than simply
that I see that it follows and do not like it. And most often there is nothing
that might even be mistaken, as an unpleasant conclusion might, for an
overwhelming force; but with or without some hesitation we simply write
or say the conclusion.

2 Another person is not predicting how it will be with me, predicting
that I will find myself quite unable to do otherwise, when in a formal
context he says that I must do this or that: he is insisting, demanding, or
requiring that I do something. The 'must' here is like the 'must' in 'You
must remove your muddy shoes when you come in to my house'; and
likewise the 'cannot' is the 'cannot' of refused permission, and is better

expressed by 'may not,' as in 'You may not eat the daisies in this garden.' Other people *will not let* us proceed just any way in formal contexts, and when we are trying hard, we will not let ourselves do so. We are not causally determined to accept a logical conclusion, but *pressed* to do so [RFM I 116].

3 Yet while these words of necessitation have the character of injunctions, it is not, as it is in the case of many other injunctions, on the authority of another person that we accept them, nor does anyone regard either himself or anyone else as free to decide *what* shall be enjoined. We want to say that something is enjoined in a formal context, not arbitrarily, or even merely *reasonably*, but because independently of the person enjoining, it is *right*. People are guided in what they insist on by the way, independently of them, ideas themselves are related [RFM I 8, 23, 119; PI §§218, 219, 352, 516].

Can we, however, give any substance to this notion that there are definite relations between ideas, with which we must so to speak fall in line, or is it merely another expression of our conviction that there is a right way to go in logical matters? Are we to say that ideas do have a life of their own, or only that it is as if they do?

If ideas literally had a life of their own, one might expect that we could watch and report on it, truthfully or otherwise. But we do not watch the gyrations of ideas; we are not fascinated or surprised by what they do; we do not make up our minds as to how much of what we observe we should report on, or in what way; we do not have to wait until a conclusion emerges before making an inference, or study it to see what conclusion it is; and it does not seem to us that we are *lying* when we make a false inference, either misrepresenting what we saw, or purporting to have seen something when in fact we saw nothing.

Of course, nobody thinks that the supposed independent life of ideas is something that we study and report on; but if not that, then what is the cash value of this notion?

Someone may say that the independent life of ideas is shown by such facts as that there are delightful surprises in store for us at the conclusion of certain proofs, and along the way too, and that we can be forced most reluctantly to recognize contradictions and paradoxes. Such things, however, show only that it really is *as if* ideas had a life of their own: they do not show that to be true.

If it is therefore false that ideas do function independently of us, and we are only saying that it is as if they did, we thereby resolve no difficulties, but merely adopt a picturesque way of expressing what is to be explained. For it would explain the incumbency that is upon us to go in a definite way if the way was prescribed by the ideas themselves; but if that is not true, we are left with the bare conviction that there is a definite way to go.

4 The conviction that there is a definite way to go is what needs explaining, and what we vainly try to explain by saying that it is as if ideas had a life of their own. This conviction is a somewhat complex attitude we all have towards mathematical and logical questions: we treat them as being definitely answerable (whether or not *we* have the right answer to them); we do not regard them as 'matters of opinion'; we do not apply the concept of probability to them as we do to some other questions; if we get different results when we perform a calculation twice we have no doubt that we have made a mistake somewhere and at least one result is wrong; we do not regard the performance of a calculation as being an experiment, and do not analyze to find the precise conditions under which it will come out the same, or will come out right [z §299; RFM V 46]. In short, we perform according to an ideal of definiteness and unmistakeability, even if we neither find our own thinking always to match the ideal, nor find anything beyond ourselves that serves as a model to which our thinking can approximate [RFM V 40].

(The word 'ideal' here may be somewhat misleading, if an ideal is thought of as something to which we strive to approximate, and the complete achievement of which is perhaps always beyond our grasp. In formal contexts nothing short of the actual achievement of the ideal will do, and it might therefore be better to call it a requirement. This may be why Wittgenstein (RFM V 40) rejected the word 'ideal.' He preferred the word 'norm'; but the 'by and large,' 'on the average' connotations of this word seem to me to make it a less-than-happy choice.)

For short, let us call this attitude the 'formalist complex.' Training people in formal procedures is in large part a matter of inducing or generating in them this complex; and it is when or because we are imbued with it that we are inclined to say that it is as if ideas had a life of their own. It *is* as if this, and other things too, were true, that is, the kind of attitude we have is just the kind we would have if such things were true. But it is only as if they were true: we do not operate in the way we would if they were true. We do not, for example, watch and record the life of ideas, or stand back and let the rules generate their own application.

5 We can see our inclination to say that we must, to be right, go such-and-such a way, as an expression of the formalist complex, of the refusal to tolerate difference of opinion, or to tolerate inexactness. On the occasions when we use words of necessitation, we think we have achieved the ideal of exactness, and that no departure from it is permissible.

We may fancy, at such junctures, that we have ascertained how certain ideas, independently of us, are related; but clearly all that is necessary to warrant our use of words of necessitation is that we should believe that we have achieved the formalist ideal. If we should later come to believe that we were mistaken in this, that will not show that we have *misused* those words. The proper occasion for their use is when, rightly or wrongly, we think we have achieved the ideal of exactness.

6 We look the wrong way upon the formalist complex: we think we have had to acquire it because of the nature of and the relationships between formal concepts; whereas I am suggesting that it is because we have the attitude that we are disposed to attribute to formal concepts that clarity and definiteness of relationship that anyone will tell you is their hallmark. We *create* that clarity and definiteness by our absolute refusal to tolerate difference of opinion in formal contexts, by drilling students remorselessly in definite ways of proceeding, and by exuding and encouraging anxiety about the correctness of results, and double-checking to be sure.

7 If what I have just said is true, there should be no problem about logical compulsion; but is it true?

The idea that we are not forced by the nature of formal concepts to cultivate the formalist complex suggests that it might be merely a cultural accident that we have this attitude and that we might quite well have developed a different attitude. That seems absurd. We want to say that it is no cultural accident that, for example, 25 x 25 = 625, no more and no less. We cannot suppose that there might quite well be people for whom 25 x 25 = 700 [PI p. 226]. We would not be prepared to say that they were just different from us, but that anyone who wanted to play it our way would have to get 625. One could make 25 x 25 come out other than 625 only by tampering with the symbols; by making = mean what we would mean by <, or making x x x mean what we would mean by x x $(x + 3)$, or by making the numerical symbols mean something different from what they mean in our system. But for any system containing any such modifications, 25 x 25 is not just *treated* as equal to 700, it *is* equal to 700. This, we want to say, is the peculiar fact that needs explaining.

8 This seems convincing; but of course it is only another expression of the formalist complex, and only shows how entrenched that attitude is. If we are looking for a justification of an attitude, we will get nowhere by reiterating expressions of it.

9 If one wants to show that the formalist complex is forced upon us by the nature of formal concepts, one must find some theory less fanciful than Platonism that shows that logic and mathematics are precise and rigorous independently of us. Platonism is the mere belief that this is so, and is unsatisfactory because it builds no bridges between the supposed independently existing ideas and the human activity of formal reasoning. If no tidings from the world of ideas actually reach us, then for practical purposes it might as well not exist.

The following are three, and no doubt not the only three, ways of bringing the supposed independent rigour of formal concepts within our practical grasp. They vary considerably in their initial plausibility, and are not offered as representing standard or stock philosophical positions.

i We might say that we begin with certain stipulations, which in the case of arithmetic would perhaps be of two kinds: (a) a set of symbols ordered in a certain way (the numerals 0 to 9), and (b) a set of operations that can be performed on or with these – a method for generating any number from these few symbols, and a set of procedures such as addition, division, multiplication. *What* system we adopt may be regarded as a question irrelevant to this issue and determined mainly by convenience and ingenuity. Any system we do adopt will generate all its propositions as it were mechanically: when we do what the procedures say in the way they say we cannot fail to get the same results.

Although it may not gibe with everyone's usage, we might call this view 'constructivism.' Its charm is that it seems to contain its own authentication: it gets us away from any crude form of Platonism by not requiring us to suppose either that the elements of the system are true (we simply adopt them), or that the propositions of the system are true if they correspond to something. They are true or correct if they are prescribed by the elements.

In spite of this, constructivism turns out to be a disguised form of Platonism. For what does 'doing what the procedures say, in the way they say' come down to? Either the procedures themselves grind out their own development and we accept the results so yielded, or it is only as if this happened. The former alternative, however, is not tenable: we do a calculation; but we cannot seriously think that, in addition to what we do, the calculation is being performed by the system itself; nor can we suppose that we ourselves are an embodiment of the system, and grind out the answer mechanically. We are too liable to error for that.

So we are bound to say that it is as if there were a machine, or that it is as if the system itself delivered the answers. But an imaginary machine will not serve to justify our computations, but only to express our conviction that there is a definite right and wrong way of managing them.

A variant of constructivism might be suggested which stresses the commitment involved in the adoption of the elements of a system. We commit ourselves to the propositions of the system in adopting the elements. This, however, obviously does not explain the connection between the elements and the propositions, but at best explains why, given that anything is a proposition of a system, we must accept it. The same question arises here as to whether the elements do, or whether it is only as if they do, generate the propositions.

ii We might say that there are materials for reasoning, and then there are the results we get in using those materials. The materials, for reasons we have already seen, cannot be supposed by themselves to produce the results, and therefore we must have some other way of satisfying ourselves of the correctness of the results. But we do have this, in the agreement of different people as to the results. The correct results are those that people agree on.

There is, of course, very broad agreement in the results we get when we reason, but this is not because we strive for agreement, but (we prefer to think) because the machinery works so well, and always produces the same results. We are not, like a children's dancing class, guided by what the next person is doing, but by the instructions, the procedures. When there is disagreement as to how something should go, the question is not settled by taking surveys to determine what the majority opinion is, but by exhibiting and re-examining the procedures by which the results have been reached [cf z §§429-31].

iii It is not clear whether the project of founding mathematics on logic is in any part intended as a solution to our problem, but should anyone propose to use it that way, saying for example that calculations can be shown to be right by logical, as well as by mathematical techniques, there will be the following difficulties: (a) There will be the same problem about the reasoning employed in the founding system as there was about that employed in the founded system. (b) The founding system will have to be so devised as to justify what we do, and in this sense will be founded on the founded system. If one devised a logical system from which a mathematics different from ours could be developed, that would be interesting, but would neither show that there was anything wrong with the mathematical system we have, nor provide a reason for preferring the new system. We would adopt the new system, not because it had a foundation that the other lacked, but only if it had some such advantage as that it was easier to teach or simpler to use. And if a logical system were devised from which just our mathematics followed, this would only show that there are more ways than one of skinning a cat.

10 We wanted to say that we do not just treat carefully examined logical inferences as being right, we are right in so treating them; and I have shown how three ways of substantiating this claim turn out to be empty. No amount of criticism of particular candidates for a justification would of course ever show that the claim was an empty one; but the following considerations may bring us a good deal nearer to having shown this:

i It is no part of any present method of formal reasoning to check (as we might put it) for truth as well as for correctness, that is, not only to reason carefully using all available checks, but also to compare what we do with some model of correct reasoning. But any general justification of what we do would require that there be a procedure of the latter kind. To imagine that the propositions of an argument and the rules of inference generate their own conclusion is vain and merely expresses our conviction as to the rigour of formal proceedings, unless we can actually observe this to happen, and use it to authenticate the inferences we make.

ii If there is or if there were a check for truth as well as for correctness – for example, if we could compare the way we reason with the way the ideas themselves function – this further procedure would be or would become part of our method of reasoning, which would itself be in the same need of justification as our other procedures.

iii If we developed some other kind of procedure for authenticating our reasoning, it would be an acceptable procedure only if it yielded the same results that we now get, and would therefore not validate our present procedures but would represent an interesting alternative way of doing what we now do. It might be as useful a further check on any particular piece of reasoning as adding a column of figures from the top down is when we have first done it from the bottom up, or multiplying to check the correctness of a division; but it would have no better claim to be the solution of our philosophical difficulties than those procedures have.

11 The conclusion that appears strongly indicated is that there is no justification of the formalist complex. That is not to say that it is wrong, or that we should abandon it. There is no particular reason to believe that only those things for which there is a justification are right [RFM V 33].

12 One might, however, wonder whether it is not justified by success – by such facts as that mathematicians do not disagree; that when we produce calculating machines that generate results just from their design components, their results agree with our results; that we can continue mathematical series beyond anything that we are trained in or agreed upon and all do it the same way; that we do discover mistakes when we check back after getting two or more different results. If it is just an attitude we have, would one not expect to find all sorts of things that do not fit?

Yet how significant is it that the formalist complex is successful in these ways? It, after all, has ways of guaranteeing its own success. In the first place, existing over the years, it creates and perpetuates its own justification. We do not *let* people reason in a slapdash way, but drill them mercilessly until they do it the way the attitude requires [RFM I 4]. And in the second place we systematically discount everything that might have counted as evidence against the success of the attitude. You are not doing this series if you do not do it this way [RFM I 4, 54, 120, 162; II 66, and many others]; anyone on whom our methods of instruction do not work is not treated as someone who reasons but reasons differently, but as someone who does not reason; and the facts that we have to work so hard to teach people logic and that we make so many mistakes in reasoning, are not treated as showing that reasoning is not independent and self-sufficient, but as showing that human beings are not up to the perfection of reasoning.

13 The upshot of the foregoing arguments is that our picture of the objects of formal proceedings as 'having a life of their own,' as existing independently of us and awaiting our discovery of them, is an empty objectification of the formalist complex. Although such Platonism never had any merit as an explanation of how it is that we are able to do logic or mathematics with such assurance and exactitude, and agreeing so remarkably in the results we achieve, it did have a soothing effect, making us disinclined to worry about the problem. We thought we knew the *form* the explanation would take, even if the substance of it was yet to be worked

out. The destruction of Platonism may therefore unsettle us, appearing to cast formal proceedings loose from their (imaginary) moorings. It may now seem as remarkable as it should have seemed all along that (as I shall generally put it) we are able to reason creatively, that is, after a comparatively small amount of training, to proceed confidently and correctly in areas not specifically covered in our training, and in some cases in areas new to the human race. Consideration of various aspects of this question will occupy most of the remainder of this essay.

14 The ease and confidence with which we proceed in regions that are new to us does seem to suggest that our, for example, mathematical training has brought about the establishment of some self-sufficient generative mechanism of the kind the constructivist envisages. For it is not that I have been trained to write 100038 as the next number in the +2 series after 100036, or that I have found it to be the case that when I do this, other people accept it. Nor am I guessing what they will say, or hoping I have done it right. But I proceed easily and with the utmost assurance; and should there be disagreement at least about such a point as *this*, I am not for a moment inclined to retreat, but am quite sure that the person who disagrees is spoofing, or has not heard what I said, or is an absolute mathematical bungler. But if, as I have argued, it is idle to suppose a mechanism of this kind, then how do we manage creative reasoning?

15 Stroud[2] explains it this way: the series up to any familiar point is like a well-trodden road that comes to an end. We see where it is leading and extend it. We know we are striking out in the right direction because it is the direction of the existing road.

No, there is no point at which it is as if we had come to the end of a well-trodden road, at which we wonder where to build from here, and conclude that such-and-such is the same direction. To the trained mathematician the path seems well-marked all the way. It *is* as if there were rails along which we travelled. But when we reflect that in spite of this really the desert is trackless, we should not say that it is as if there was a road to the edge of the desert and we extended it. It is not even *as if* that is the case.

16 Some of the puzzlement as to how creative reasoning is possible may arise from the fact that we are apt to say that we *know* how to do it. In the sense that we do it readily and correctly, that is true; but the appearance of the word 'know' here may mislead us, if it is treated as an *explanation* of

2 'Wittgenstein and Logical Necessity,' *Philosophical Review*, LXXIV, 1965, 518

our doing it readily and correctly. Then it suggests that we have some information or instructions that steer us unerringly.

17 Is it necessary to suppose that in some such sense as this we know how to proceed? Well, all acorns that grow, grow into oak trees, and none into pine trees; and they do not know how they are to develop, they just develop that way. Having knowledge is not the only way of proceeding unerringly. But what of that? People's logical performances are in many ways unlike the growth of plants. People do not reason creatively without instruction; and what we have been taught we surely know.

Suppose, however, that there was a nut that when tapped six times would grow into an oak tree, five times into a walnut tree, and when not tapped would not grow. How it grew would then depend on something outside itself, the way our logical performances depend on instruction; but we would not be inclined to say that when tapped six times it *knew* it was to grow into an oak tree.

The tapping in this case still seems too unlike instruction, however. We feel that the oak tree potentiality was not supplied by the tapping, but must already have been contained in the nut, only to be released by the tapping, whereas to be like instruction we would want the nature of the development to be contained in the tapping. So let us take a more complicated case.

Suppose there was a kind of nut about which all we knew was that tapping it affected how it grew. There was no discoverable rule about the effect of various tappings; but we could tap it and plant it, and when we saw how it was developing tap it some more until it showed signs of having become the species we wanted. It would not be the case that this nut contained in itself various potential ways of developing, but rather it would develop into this or that entirely as a result of outside interference. The tapping can now be seen as artful, as instruction is, and like instruction, as intimately determining the resultant development. But still we would not say that the nut had been taught, and knew, what to be. We just tap it until it promises to grow as we wish, and then it does – or perhaps it does not [PI §145].

Of course, teaching a person is a much more explicit process than tapping a nut, but we are in one way like the imaginary nuts: we are taught until we show clear signs of shaping up in a certain way, and then we do shape up that way – or we do not, and then we are taught some more, or abandoned. And we will see later on (§§42–4) that our teaching procedures are in many ways less explicit than we are inclined to imagine, and more like tappings.

18 There is, perhaps, an inclination to say at this juncture that although it may not be *necessary* to suppose that there is some information or instructions that we have that show us how to proceed, still in fact it is the case that there is: we learn rules, and they show us how to go. What is the objection to saying that it is our use of rules that explains our ability to reason?

19 It generally seems very clear to us – unmistakeably clear, one might say, in the case of well-stated rules – how a rule is to be applied. Do we really not know just from what rules say, how to follow them?

We see the rule and then proceed very confidently; but surely this is because we have already learned how to follow it; and it is certainly not clear that it is just the rule, rather than the rule together with our training in following it, that determines what we do.

Moreover, our training in following rules cannot consist in further rules as to the interpretation of a given rule, or we would find ourselves in an infinite regress. While we may at times suggest other wordings of a rule to help this or that person, essentially the training must simply be a matter of getting us to proceed in a certain way, given a certain rule [PI §201].

20 Furthermore, the role of rules in our thinking is often greatly exaggerated. It is an interesting philosophical problem how we are able to follow a rule; but it is not a problem because all thinking is rule-following or rule-governed, but only because there is such a thing as being given a rule, and following it.[3] Our thinking may be regular, that is, of such a kind that it is possible for an ingenious person to formulate a manageable set of rules by following which one could think in just the way we do; but it does not follow from this that it is by following those rules that we do think.

Nor is it, in any straightforward sense at least, the case. We do not *think* of the rules as we proceed; most of us could not say what rules we follow; and if anyone does say, it is not clear whether what he offers will be the rule he has in fact employed, rather than a rule which, had he employed it, would have led him the way he went. (We are generally prepared to discuss whether such-and-such was our rule, and to make changes if it appears that a candidate will not quite account for what we have done; but

3 A formula for developing a series is an example of such a rule. The formula 2n + 1 expresses the rule: develop this series by multiplying any given member of the series by 2 and adding 1.

it should be a matter for avowal and not for discussion, whether we did in fact employ a certain rule.) [PI §82]

21 Confining ourselves, however, to the question how we follow a rule on the occasions when we (undeniably) do that, it may yet appear to be a difficulty that we are not trained in the following of each and every rule, but that with a certain amount of training in rule-following, we are able to follow rules that are new to us, and do it without difficulty and correctly.

We do, however, also receive training in rule-following in general: we are drilled as to what, in the application of a rule, to change, and what to keep the same, when the rule is changed in various ways – until it comes about that when we are given a new rule, without further advice we know what to do.

And here again, the training does not so fix us that we will never go wrong, but we reach a stage where it generally seems clear to us what to do and we do it, and where generally we are right. But we can go wrong, and then perhaps we are drilled some more. And it is not treated as logically impossible that even the most carefully drawn analogies with similar rules should lead us wrong. We are prepared to be shown an intelligible sense of a rule different from what we have taken it to have – although we are generally inclined to insist that it should be understandable how it should have this sense. Something is wrong as long as it appears quite arbitrary that a rule should have the sense it has.

22 We seem on the one hand to want to deny that the rule itself prescribes its application (§19), but on the other hand it seems wrong to say that a rule is just a proper name for a procedure, having no internal connections with it – as if we could not 'get the hang' of rule-following, but simply had to learn what people do in each and every case. We demand to *understand* rules – for example (§21), to be shown how it is that a certain rule applies the way it does. But is this not as much as to say that rules do prescribe their application? Can we have it both ways?

23 This difficulty may arise from the fact that there are so to speak two levels from which the matter is being viewed – that the levels must be kept distinct and that we have been conflating them. There is on the one hand the learning level, and on the other the mastery level. At the learning level a rule initially means nothing to us, and we must simply do what we are told by way of applying it. But the teaching is supposed to illustrate what the rule means, and to bring it about that from the rule itself we will see how to develop it; and further, to bring it about that we see in general how

the terms of rules prescribe applications. If I teach a person to bring the object on the left when the bell rings, but then have to start all over again teaching him to bring the object on the right when the whistle blows, then he does not yet understand rules, but a rule is for him a complicated noise that serves as a signal to do something. Training in the use of rules is not just a training in doing things in a certain way, but also, queer as this may sound, a training in why this is the way to do it.

Hence, from the learning level to the mastery level there is a transition from doing as one is told to doing what the rule prescribes – from external to internal guidance, as one might put it. When a person has achieved mastery it seems to him that the rules themselves direct him, and that the agreement that exists as to how they should be applied is an accident that is only to be expected since the rules so clearly prescribe their application.

24 At the mastery level one cannot say 'The rule does not guide me.' That is as much as to say that the training is not yet successful. And if one says it and also follows the rule correctly, and given an alteration of the rule makes appropriate modifications in the way one follows it, it will be either false or unclear what one means by saying that the rule does not guide.

But of course when we denied that the rule guides, we did not mean to deny that given it, we know what to do. *That* is consistent with the supposition that the rule is the rule it is because one has been trained to follow it that way. What we meant to deny is that it is conceivable that without any training we should see how the rule should be followed – as if it were only a human imperfection that we so universally need training – gods would understand right away. A strange conception, yet that is how it seems to us when we have reached the mastery level, because it is so clear to us just from the terms of a rule what we should do.

At the mastery level we can say no other than that the rule guides; but that does not mean that the teaching has brought it home to our slow mentalities how the rule of itself prescribes its application, has shown us the logical mechanism that works that way whether we grasp it or not. It only means that given the rule, we know right away how to follow it, and that we have a measure of self-confidence or independence about our understanding of rules in general – for example, we will not be bashful about defending our interpretation of an unfamiliar rule, and we will be leery of an interpretation of a rule than cannot be justified in a familiar way.

25 Although we cannot deny that in a sense rules guide us, prescribe their application, this is just how it does and should look to us when our

training is successful. But we go wrong if we take this to mean that a rule has a certain application independently of human beings, and whether we like it or not. It is in fact the rules plus the training we have in their use that determines what we do.

This, however, seems to cast us loose again from our moorings. Training seems too uncertain a process; and the only check on the (uncertain) training of one person that we now have is the (uncertain) training of others. Our problem about creative reasoning reasserts itself: how is it that we are able to get the hang of it, become self-propelled and able to proceed readily and correctly in areas not specifically covered by our training?

We feel that this question would be nicely answered by the idea of a logical mechanism, with a comparatively few components working together to generate endless developments. We understand how a machine can do this sort of thing, and how, barring defects in materials and workmanship, various machines will generate the same developments [PI §193]. But if, as we have been saying, the training does not create that kind of a mechanism, independent thinking becomes painfully difficult to explain.

26 Well, why does it? What needs explaining here? It is the most normal thing for people to start thinking independently after a comparatively small amount of instruction. It needs explaining when this fails to happen. Then we perhaps find brain damage, or find that there are some people for whom new techniques of instruction must be devised. But one explanation of a person's ability to think independently is that he is normal, and has had the usual training.

27 What may stop us is the idea that training can only be a matter of learning by rote, and that on a rote-learning basis our abilities can reach only as far as our training, and we can never become self-propelled. Yet, what makes us so sure that this is the case, when we see training having a different effect all the time?

28 One might reply that it is not so certain that it is *training* exactly that has this effect; or if it is training, it is not so certain that in the course of it some internal chemistry does not bring about the establishment of a generative mechanism.

This, of course, only expresses a conviction that there are only two possibilities: either a device does only what it is specifically set up to do, or it is a generative device, with components enabling it to develop behaviour endlessly. (It might be suggested that we think these the only alternatives because that is how it is with machines. But that does not really matter.)

29 If we can see no way between these alternatives, what would we say if some mad tinkerer produced a machine that in some fairly literal sense had to be taught how to function: it could do nothing with rules until shown their application, and would initially reproduce only the applications that had already been illustrated, but it had a button marked 'and so on,' the pressing of which would result in its going beyond its instruction, sometimes the way we would like it to go and sometimes not. Further, when it went on wrongly, it could be shown the correct development, and then would generally revise its mode of operation at least up to the point that had so far been illustrated to it; but then when the 'and so on' button was again pressed it would sometimes go on as we wish, and other times not; but whenever it once went on correctly, it would thereafter generally go on in the same way when it reached the same juncture.

We can imagine that the tinkerer had no idea why the machine behaved this way: it was just an attribute of one of his creations that he had encountered by chance. But he could produce replicas of the device, and he found that some of the replicas were as we might say 'quick to learn,' others would eventually learn, and still others would never do anything right when the 'and so on' button was pressed. These he would scrap; the others he would keep and do with them what he could.

We might be astonished that there should be a machine that behaved in this way, and try every way to show how it was a variant of one of the known designs of machine. But would our astonishment not merely be a product of our insistence that somehow it must be one of the known designs? There would remain nothing surprising once we accepted that it was simply a new kind of device, and that it was pointless to try to decide which of the old types it was.

Nor could it be said that the new device was unintelligible. We could study it, and learn how its operation is affected by cutting this wire, adding this circuit, pressing the 'and so on' button at unlikely junctures. We could find what kind of care and maintenance it required, how its durability and efficiency is affected by being made of various kinds of materials; and in such ways achieve as much understanding of it as we now have of, say, a watch mechanism. We understand a watch mechanism even if we do not know how it is that the spring exerts the pressure it does. It is enough that it anyway does this, together perhaps with the fact that we know how springs of different materials or thicknesses or lengths affect the time-keeping capacity of the watch.

The point of this analogy is to suggest that our difficulty in understanding how people reason creatively may arise in part from an inclination to insist that this phenomenon must be reducible to some known model of explanation, and that if we could regard people, as it was suggested we might regard the mad tinkerer's device, simply as a new kind of mechanism, there would either be no problem, or not that problem. It should not after all be so very surprising that people are unlike machines.

How people do it would not then be unintelligible. We could study them the way we studied the tinkerer's creation, and perhaps discover what kinds of difference between one brain and another made for quick or slow learning, for inventiveness, for proneness to error. We could discover how cutting out or even adding brain tissues affected people's performances, and we could certainly discover, as we already have, the differences made by different techniques of instruction, or by drugs, alcohol, or vitamins.

30 'How is he *not* able to do it?' would then be explained, as it in fact often is, by saying, for example, that he is drunk, he is only a child, or he lacks a certain piece of brain tissue. And 'How is he able to do it?' would be explained by saying that he is normal, and has had the usual instruction.

Although these two sorts of explanation are obviously correlative, the former seems all right to us, while the latter is disappointing, not at all what we had in mind, a piece of philosopher's mischief.

31 What did we have in mind then? Perhaps it is as if a magician had performed some remarkable feat, and when I enquired how he did it you said 'He is an experienced professional. He is not just a beginner!' I would feel like replying 'I know that. What I wanted to know was how experienced professionals do it: what their system is, what techniques or procedures they use to yield that result.'

32 Yet can this be right? For in the case of reasoning I do it myself, I know all about it. The magician can do something that I cannot do, and I want to find out how he does it, perhaps so that I can do it myself. If my question about reasoning were really parallel, it would be an answer to say 'Come along to my classes: Tuesdays and Thursdays at 11' [PI §369]. If anyone does not know how to do it, the remedy is instruction.

Nor is there any particular problem to finding out how a person does it – for example, if he calculates with amazing rapidity. We ask him, and he perhaps tells us some of the shortcuts he employs, and then we know just how he does it.

33 If the problem is hence not one of finding out how we do it, it may appear to be that no matter how fully we appreciate what happens – how we do it – what we therein know does not seem to explain the results we get.

When we reflect on the conscious processes of reasoning it can sometimes look as if the mind is a weird machine from which most of the connecting levers and belts have been removed, which, however, amazingly manages to function in just the way a complete version of such a machine would, and which furthermore has a remarkable capacity to materialize appropriate parts as if by magic, which parts in turn have their effect without always being connected with the mechanism [cf PI §§635, 653].

How, for example, did I know, when asked to multiply 19 x 5, to multiply 20 x 5 and subtract 5? Well, I have done that kind of thing before, and found it a convenient shortcut. But that is not the kind of explanation we want. It is like saying that I am normal and have some training and experience. We want to know how that piece of the machine materialized just then, when it was needed. And it is for all we can see as if it had appeared by magic.

How did I know that 20 x 5 = 100? Well, I remember it. But that is just to say that without hesitation I put 100 as the answer, and that I am presumed to be able to do it on account of past experience. But that does not explain in particular how that information came back to me right then.

34 Suppose we had some kind of an answer to such questions: perhaps an account of how traces are left in our brain tissue by past experiences, and activated now by certain stimulations. I think we might still feel that we had not accounted for the *intelligence* exhibited in independent thinking. For not every move we make is a repetition of something we have done before; and certainly not every combination of moves, every strategy. It almost looks as if the only thing that would fill the bill is to suppose that the connections in the weird machine are supplied by a little mathematician. He devises the strategies and does the reckonings that, coming to us often in a completed form, appear simply to have been remembered.

But we are ourselves such intelligent beings; and if we were to explain our own intelligence by means of some other intelligence, that other intelligence would remain unexplained.

The difficulty seems again to be that we cannot accept our uniqueness, but keep trying to reduce it to something else (cf §29).

35 It may seem plausible to suppose that our difficulty in explaining how we reason derives partly at least from the fact that there are so many

different ways of doing it. We keep looking for one system, whereas in fact there are indefinitely many. One person may, for example, have memorized the nineteen times table up to a certain point, while another uses some stock multiplication method, and others have various ingenious shortcuts. Everyone has his system, but not everyone has the same system. But what makes us so sure that everyone has his system, – that a working system is the form of explanation required here?

36 Chiefly two things: (i) we reckon that we could understand creative reasoning if we could suppose that in every case it was generated by some system. It is characteristic of a system to be capable of endless developments from limited components; and part of what we do not understand about creative reasoning is how, after a comparatively small amount of instruction, we become self-propelled. (ii) People do in fact have systems. If you ask them what their system is they will nearly always tell you.

37 But how much explanatory power does the idea of a system have, given the kind of instantiation of it that can be attributed to human beings? It does not lack this power in the case of machines; but with a machine we can see all the parts, and how they are related to one another, and exactly what it is that makes this part come into play in certain conditions and not come into play in similar but still different conditions, and exactly how a part achieves what it achieves when it does come into play. In its human instantiation, however (§33), the system seems to lie mostly beneath the surface, and to elude us. We see only one or two parts of it at a time, and when we try to press for a complete specification of a person's system, at every stage it seems necessary to suppose that there are not yet specified parts connecting up the parts so far specified; and the further we go in probing to the depths of the system, the less believable it is that we *use* this cumbersome mechanism and the more it seems as if we are inventing a system to satisfy our demand that there should be an explanation of the system type. To the person who is being cross-examined, it seems less and less as if he is remembering how he does it, and more and more as if he is working out how it *might be done*.

38 For it is very unclear whether a person's answers to our questions as to how he reasons will be actual reports of what he has done, or inventions – statements of what he might have done had he followed some method. There are a few clear cases, such as my method of multiplying 19 x 5. They are characterized perhaps by a definite recollection of just now having done it that way. But under this kind of questioning our

recollection is not usually so clear. Is this because the procedures are so familiar that we hardly notice our use of them, and because, not often being required to explain these things, we are not in the habit of taking note of and remembering them? Or is it because there is nothing to be remembered? Perhaps we just do a calculation, and afterwards are sometimes able to produce a justification of it. The justification shows that it is right, but does not show how we did it.

Certainly that sometimes happens. Something seems right to us and we do it. When a doubt is raised as to its correctness we are surprised, but soon it occurs to us that what we have done ought to be right because it is so like what we have been taught to do in such-and-such a very similar case, or because it looks like a perfectly routine application of such-and-such a rule. But we need not have been mindful of such facts when we acted.

Does everybody have his system? Well, different people will do a calculation differently; but if we call this 'having different systems,' we are not using the word 'system' the way we use it when we suppose that it would explain creative reasoning if everyone had a system.

39 We may be led to think that we possess at least a rudimentary knowledge of the human design that enables us to think by the fact that we say such things as 'I suspect you were misled by such-and-such an analogy. The following analogy would be better ...' This looks like a hypothesis as to how this person functions, and an attempt to substitute one mechanism for another. But (a) it is not clear that in saying such things we have a theoretical interest in accounting for what has happened. If our explanation is rejected, we are not likely to persist until we find the correct one, or else go away puzzled and frustrated. Rather (b) we make such remarks for heuristic purposes. We want to say that such-and-such anyway is a misleading analogy, whether or not the other person was in fact guided by it, and further that the contrast between it and what we offer as a better analogy may serve to bring out something important. We do not offer the new analogy as an infallible guide, as a part for a properly designed mechanism.

40 We use such remarks for instructional purposes; but if they have their effect, must we suppose that something in them has become incorporated in the mechanism, and henceforward contributes to its working the way it does?

As a result of what we say, a person may no longer make the mistake he has been making. But people are not always mindful of the instruction they have been given when they proceed correctly, nor do they necessarily or even generally reproduce parts of that instruction when they in turn are teaching someone else.

If the substance of the teaching became part of the mechanism, one might expect that in teaching we would be trying to produce the same mechanism in other people that has been produced in us, and that teaching would involve becoming aware of the structure of one's own mechanism. Differences in effectiveness between different teachers would be explained in large part at least by differences in the clarity of their self-consciousness.

In teaching, however, we are not showing how our own mechanism works. Rather, it occurs to us that such-and-such an analogy may be useful at this juncture to the person who has made this mistake. We teach ingeniously.

41 Teaching is more like adjusting a mechanism than manufacturing one, although it is not very like either one. We do not install parts, but we say things that have an effect on the way people operate. We tune the engine until it runs smoothly and then it does. Or it does not, and then we tune it some more.

42 If you are inclined to think of teaching on the model of designing and installing parts in a mechanism, consider the following: suppose it became noticeable in the schools that there were more and more children who were unable to learn mathematics. It was not that they were disaffected, or did not try; but try as they would they continued to make blunders, and were unable to see where they had gone wrong when their mistakes were pointed out to them. The efforts of the best conventional mathematics instructors failed; but a man appeared who claimed he could teach these children. As lessons proceeded in the classroom he watched the problem students, and with the air of someone who knows what he is doing, played here a note, there a phrase, on his piccolo. The effect was not magical: some students still went from one mistake to another under his influence, and it took varying lengths of time with different children before improvement was noticeable. But in time many of the students learned their lessons, at least to the point of making no more mistakes than their normal classmates. Other people tried the piccolo treatment on their problem students, but seemed never to understand what to play when, and it did not work.

Here (artfully contrived) piccolo playing was essential to the success of the instructional process, and we could even imagine this being so if there was no concurrent conventional instruction. But we would not be inclined to say that piccolo tunes became part of the mathematical mechanism. Musical phrases simply could not serve as mathematical guides.

43 What bearing does this little fantasy have? What is there in regular mathematical instruction that is analogous to piccolo playing? We tend to think that everything that is said could (as we put it) be incorpora.ed in the mechanism; but there is really a great deal, when you think of it (when you pay attention to the patter [RFM III 27]), that is not of this kind: not only expressions of rejection, agreement, encouragement [PI §208], but particular emphases, particular ways of writing things down, more or less far-fetched analogies – anything that may help someone to see how it goes. We perhaps teach the number series by writing it down this way:

$$
\begin{array}{cccccccccc}
 & 0 & 1 & 2 & 3 & 4 & 5 & 6 & 7 & 8 & 9 \\
1 & 10 & 11 & 12 & 13 & 14 & 15 & 16 & 17 & 18 & 19 \\
2 & 20 & 21 & \dots \\
3 & \dots \\
10 & 100 & 101 & 102 & \dots
\end{array}
$$

and drawing the student's attention to the way the basic sequence 0 to 9 is repeated in each line, first with a 1 in front of each figure, then with a 2. We might do this by underlining as we talked, or by connecting loops between the 1s, 2s that appear underneath each other and horizontally along the same row, and in many other ways [PI §§143, 145]. But we are not teaching the student that he should write the numbers in rows in this way, or that he should underline them as we do or connect them with loops or anything else. Such things are no part of what he *has* when he understands, but only serve (or do not serve) here and now to bring it about that he does it the right way. They are left behind once that result is achieved.

44 Such tactics are unlike piccolo playing in that they seem to us well calculated to produce the desired result. But on the other hand we did say that the piccolo player was proceeding artfully: and we can quite imagine a breed of men who found it just as incomprehensible why we should carry on the way we do, with encouraging smiles and emphatic gestures, and that it should actually work as we do the antics of the piccolo artist.

45 The foregoing seems to me to explain something that I think has not been understood about Wittgenstein's example [PI §185] of the person who naturally takes the gesture of pointing in the direction from fingertip

to shoulder. The point of the example is to get us thinking of the strange and miscellaneous things we would do to get him to understand. For we would not give up on such a person, even though with him no amount of further pointing will get his directions reversed; but (supposing that we could not explain verbally what was meant) we would perhaps draw a picture of someone pointing, with an object at either end of the line of pointing, and then erase or cross out the object that is at the end of a line from fingertip to shoulder, and emphasize the other object, perhaps circling it heavily or making a certain kind of gesture towards it. This might or might not work, and if it did not we might get a third person to enact a scene in which he pointed to an object and when we picked it up he made profuse expressions of approval. Or again we might point to an object, and then keeping our arm always aimed at it go over to it and pick it up. We would of course also get him to play, correct him when he went wrong encourage him when he went right, and repeat the game until he went right regularly. And many other things, as various and as desperate as the ways of showing a fly the way out of a bottle.

The philosophical point of noticing these things is not, I think, to bring out how pointing is tied in with other actions and gestures, but to illustrate how many curious and different things we may do to bring about understanding, and how inessentially the things we do are connected with what we are trying to teach. For we would not expect a picture of a man pointing, with one object crossed out and another circled, to become part of a person's equipment, enabling him to point as we do, but only that he should anyway point as we do.

'Someone,' Wittgenstein says [PI §590], 'might learn to understand the meaning of the expression "seriously *meaning* what one says" by means of a gesture of pointing at the heart. But now we must ask: "How does it come out that he has learnt it?" '

Here one could waste a lot of time trying to say how it might 'come out that he has learnt it'; but presumably the interesting thing is how it does *not* come out: not by his pointing at the heart. That would leave it quite undecided whether he had understood. What people get from our instruction can be very indirectly connected with the instruction itself.

46 One may still feel strongly moved to say that although the things we say in teaching may be very miscellaneous and curious, and may not themselves be such as to become incorporated in the system, still they surely somehow have the effect of creating or revising a system. The arithmetical

learner, for example, comes away from the instruction with a system for generating numbers.

That will seem stunningly clear if we are satisfied anyway that the only possible explanation of a person's ability to generate numbers lies in his possession of a system for doing it.

But now if we ask ourselves *what* system, it will seem extraordinarily hard to say. What we find to be true of a person who knows his numbers is mainly, first, that he can readily rattle them off in proper sequence; second, that he will often be able to teach someone else how it is done (but only in the miscellaneous ways we have described); and third, that if or when he also has some training in philosophy or in mechanics, he may also be able to apply himself successfully to projects of devising a formal statement of a way of doing it or of building a machine that will do it. All these things, however, seem to be expressions of or effects of his having a system: they are not the system itself. He does not, for example, employ his formal statement of a way of doing it when he does it. That is something that did not exist until he studied philosophy; and, moreover, he worked over and revised the statement until it was such as to yield the results he knows without it to be right. He did not come upon the statement by introspection.

In this way understanding comes to look like a very queer phenomenon. On the one hand we think something must be there, yet not only is it not in evidence, but anything we succeed in unearthing turns out not to be it. It looks divine: the unknowable source of many marvellous things. And yet another day it will not seem at all surprising that a person should be a capable mathematician: he is a bright fellow, and has been studying it for quite a while under good instructors.

47 The tendency of much of what we have been saying is that there is either no problem, or no problem that cannot be handled in quite familiar ways, or not the problem we thought there was, about how we reason creatively. If a person has performed some feat of independent thinking and we ask him how he did it, he will generally be able to set forth some or all of the steps he has taken. That will be a kind of answer, but not the kind we want; and we may express what we still do not understand by saying 'But how did you do *that*? What led you to take all those steps in that order, and how did you manage to avoid the pitfalls and blind alleys?' To this many people could only answer that they do not know, they just find themselves doing it that way, and of them we could perhaps only say that they did it that way because they were intelligent beings with some training

and experience in such matters. But others, perhaps with a more lively interest in themselves, might be able to say such things as that they applied strategy because the problem was so very similar to one they had once seen solved in which the strategy had worked. In saying this they might be reporting their thought-processes, that is, reporting that they had in fact been struck by the similarity of the two problems, and remembered the strategy, or they might only now remember the similarity of the problem, and be conjecturing that it had had its effect. But the explanations of the most self-conscious would soon come to an end, and regardless of where that juncture occurred, we would be left saying some such things as that they could do it because they were intelligent and trained.

That, however, seems most unsatisfying. It is like offering as an adequate explanation of the power of a piece of metal to attract another the fact that it is iron, and had been stroked with a permanent magnet. We still want to know what difference the stroking makes to the metal, and how being different in that way has that effect. And, similarly, no matter how much it is emphasized how like stroking teaching is, we want to know in the case of people learning to reason what differences in them the stroking has brought about, and how those differences enable them to (for example) solve mathematical problems. And (§32) we do not or we ought not to want to know how they go about it when they are trained, for we do it ourselves and in a perfectly good sense we know how to go about it. And if we do not know we can soon find out.

48 Suppose that by some means we were able to get a much more exact view of the living human nervous system than is now available to us, perhaps by having electrodes connecting a human brain to a large scale model in such a way that the condition and the changes in the nervous system would be accurately reflected in the model; and with the aid of this equipment we were able to discern characteristic states of the nervous system that existed when people could, for instance, multiply and did not exist when they could not, and were able further to see what slight differences in these states made for unreliability in multiplying, and by watching when various external influences were applied could see what things tended to remove the differences and what to accentuate them, and amongst the former, compare the permanency of the removing effect of various influences. We could also perhaps discover the effects of different structures (as distinct from states) of nervous tissue on mathematical ability; and if comparable abilities were sometimes shown by people with

different structures, what differences in the states were required by differences in structure. And many other things. In these ways after a time we would come to know a great deal about how, in one sense at least, we are able to reason.

49 Whether or not such discoveries are possible, the question is, if they were made, would they be the answer we are looking for? Two considerations suggest quite strongly that they would not be:

i We are not all incipient or frustrated neurologists. We are quite capable, without having any interest in or knowledge of neurology, of finding it terribly puzzling how we reason. It is not after applying to a neurologist and finding that he had nothing to tell us that we are puzzled, but we are puzzled and we would scarcely think of asking his advice.

ii We would surely reach the same kind of rock bottom in neurology that we reach elsewhere: the point where we can only say that it is simply the way it is that when such and such conditions obtain, people can do it. (Compare 'He is intelligent and well-trained.') For suppose that we not only had a general understanding of the workings of brain tissues in formal affairs, but had such a complete knowledge of a certain person's brain tissue as he did a calculation that we could say that its structures and states were of the right kind, and that nothing analogous to a mechanical calculator's slipping a cog had occurred in the course of his doing this calculation: if he got what we would otherwise say was the wrong result, no amount of double-checking as to whether his brain tissue was functioning suitably would convince us that he was right; but if we did become convinced of that, it would be by doing and double-checking the calculation ourselves.

Let me now summarize the proceedings of this essay. It was suggested that problems about logical compulsion do not arise while we are engaged in logic or mathematics, but when we stop and think philosophically about those activities. The fact about them upon which we have focussed is that we regard ourselves and other people as being bound in these contexts to go in certain very definite ways. This incumbency is typically expressed using what we have called 'words of necessitation': 'must,' 'have to,' 'can only,' 'cannot,' and so on. It was argued that these words do not describe or report anything like a causal necessity, but are rather *injunctions*: we press ourselves and other people to behave in these ways. This conclusion did not, however, explain upon what authority such injunctions are based.

We wanted to say that we so enjoin because, independently of us, what is enjoined is *right*. However, when we try to give substance to this claim, we quickly see that Platonism, the theory that formal concepts have a life of their own, with which we must fall in line, has no cash value, and must be regarded merely as a picturesque expression of the compulsion we believe ourselves to be under in formal matters.

It was then suggested that this inexorability in fact lies very close to hand, in an attitude that is most diligently cultivated in students of formal disciplines, an attitude that we dubbed 'the formalist complex.' We drill students remorselessly in following certain procedures exactly, in double-checking results, in refusing to allow disagreement, sloppiness, and guess-work. It was suggested that this attitude is not necessitated by, but rather *creates* the rigour and exactness of formal concepts.

There was then a criticism of various possible ways of showing that on the contrary we are forced by the nature of formal concepts to cultivate the formalist complex; but even given the soundness of these criticisms, there still seemed a problem as to how, as we put it, we can 'reason creatively,' that is, after a comparatively small amount of instruction, proceed independently and competently in contexts new to us. Depending on just how this question is interpreted, various answers, as we have seen, can be given:

1 We can give an account of the steps we have taken. ('How did you do it so fast?' 'I multiplied 5 x 20 and subtracted 5.')
2 We can give an account of the reflections that led us to adopt the strategy we employed. ('It struck me that the problem was similar to such-and-such a problem, and that if I allowed for such-and-such a difference, the same strategy would apply.')
3 We can offer common sense psychological conjectures. ('I really don't know why I did it that way, but I suppose it must have something to do with the similarity between this problem and the problem of such-and-such, with which I am quite familiar.')
4 We can reject the question ('Why ever not? I am intelligent and have had good instruction. It would need to be explained only if I *couldn't* do it.')
5 We might one day be able to give a neurological explanation, showing that the necessary brain tissues were there, and that they were in the states necessary for the performance of these functions.

If interpreted in any of these ways, our question about creative reasoning is readily answerable, but is also shown not to be a philosophical question. The object of parading these answers has not been to reduce the philosophical problem to any of them, but rather to bring the worry into focus by making clear what may not have been clear initially: that it cannot be handled in such straightforward ways.

What puzzles us about creative reasoning is its precision – the fact that, while there is a great deal of latitude as to how a plant may grow or as to how a person may believe in most other areas of human activity, in formal proceedings there is only one right answer. How does it come about that so many of us can satisfy this exacting requirement?

This question is made peculiarly difficult by the fact that, if no form of Platonism is true, on the one hand it looks as if the only supposition that would account for our reasoning ability is the supposition that there is at work a generative mechanism of the kind we have in a calculating machine; while on the other hand there is adequate reason to suppose that there is no such mechanism. Human beings are too unlike machines and too prone to error; we are not conscious of such a mechanism; we do not proceed in the orderly, efficient way that could be expected if such a mechanism existed in us; and our teaching methods are not at all of the kind one would expect if their object was to create such a mechanism.

It would represent a considerable advance if the considerations we have adduced to the effect that the explanation cannot lie in the supposition of a generative mechanism served to close the door on that line of investigation. Granting this conclusion, however, our problem will only seem the more acute. As long as we believed we knew the general form the solution would take, it did not seem deeply worrisome that there were many internal difficulties to be ironed out; but when we take away the prospect of this form of solution it may appear that there is no remaining direction in which to turn.

Even in this acute form, however, I suggest that the difficulty has been relieved, if not removed, chiefly by the following considerations:

1 It has been suggested (§§4–6, 24) that the precision of reasoning is not, as it must be treated in order to generate the problem in its acute form, a fact to be explained, but an ideal to be achieved. We *work at* achieving precision, will not *allow* roughness, sloppiness, guesswork. If we most often achieve precision, that is adequately explained by the strenuous and meticulous care that we devote to that end.

2 In the same connection it has been suggested (§24) that we must of course say that reasoning is exact. To deny it would be to show that a crucial element in our training had not taken effect. But we mistake the import of so saying if instead of seeing it as the expression of an attitude, a policy, or an aspiration, we treat it as a description of the way we do reason. Translated into human performance it comes, not to the sure, untroubled progress towards a conclusion that we see perhaps in the proof of a theorem, but to nervousness, head-scratching, double-checking, unwillingness to accept solutions on authority, insistence on being shown how it is that this is the right answer. And it is no great problem how it is that we perform in these ways.

3 We have repeatedly (§§17, 21, 29, 41) had occasion to make remarks like the following: 'We tap the nuts until they promise to grow as we wish, and then they do. Or perhaps they do not, and then we tap them some more.' The point of so saying has been to suggest that our problem may have been mis-stated or misconceived in another way, namely in being conceived as the problem of how it is that after a comparatively small amount of instruction people are able to proceed independently, with all the sureness and precision that we attribute to sound performance in reasoning. We may be thinking that at some point they have *got it*, the machinery is complete, and we wonder *what* they have got. But not only is it not the case (see (2) above) that people reach a point where they can proceed directly, infallibly and with untroubled assurance: they reach no point where their equipment is complete, where they have nothing further to learn, where they will no longer ever make mistakes, where they never have to show their results to other people and see if they agree. We do not at some point *know* how to go on, but rather at some point we acquire a certain independence. We no longer fear to branch out a little on our own, we feel ourselves able to discuss things on equal terms with our instructors, we can satisfy ourselves generally whether we have got something right, and would not think of looking it up in the back of the book. Again, it is no great problem how this state of affairs is possible.

4 Quite apart from the question *what* it is that has to be explained here, it was suggested that the problem may have arisen partly from being unduly cramped in our conceptions of what will serve as an explanation. In this regard our problem seemed to have the form: 'There are only such-and-such forms of explanation available. In this case it can't be any of these, and therefore it must be this. And yet ...'

No matter how much we bring our ability to reason down to earth, it can still seem to need explaining; and if we insist on an unsuitable form of explanation, we will indeed have a problem.

What I have suggested here (§§29, 34) is just that we should treat human reasoning ability as a special phenomenon, not necessarily reducible to anything else, and then it will not seem astonishing, or no more astonishing than the spring of a watch or the hardness of a table top. Nor will it be an ultimate mystery; but we will be able to study and explain it in various ways, for various purposes, mainly in ways (1) to (5) above (p. 199).

One may still feel that a problem remains until such explaining has been done; but if that is so, as I suppose it is, it is not a philosophical problem. The role of philosophy in these matters could perhaps be explained this way: good-heartedly tackling a problem in logic or mathematics is like proceeding along a familiar path. The path has its hazards and obstacles, but we are equipped to cope with these, and as long as we stay on the path, our difficulties are all manageable. But there are woods along the edge of the path, and trails crossing it and leading off in various directions. These correspond to the problems of this essay. If we allow our attention to stray from the difficulties internal to logic and mathematics, we may be puzzled by the surrounding territory, because there are no signs pointing the way. What a philosopher does is tack up notices indicating what you will find if you go this way or that, and which paths lead nowhere. He may then join the human throng along some of the paths; but it is not as a philosopher that he does so. The philosophical job is done when the paths are properly marked, and the choices made clear. The hard thing is to know when to stop [z §314]. But the difficulty here *should* be one of deciding how much to write on the notices, not one of deciding whether to proceed along any of the paths.

This book

was designed by

ANTJE LINGNER

under the direction of

ALLAN FLEMING

and was printed by

University of

Toronto

Press